S⊤ RAI **Medical entomology for students**

Medical entomology for students

Second edition

Mike W. Service

Emeritus Professor of Medical Entomology,
Liverpool School of Tropical Medicine, Liverpool, UK

CAMBRIDGE
UNIVERSITY PRESS

PUBLISHED BY THE PRESS SYNDICATE OF THE UNIVERSITY OF CAMBRIDGE
The Pitt Building, Trumpington Street, Cambridge, United Kingdom

CAMBRIDGE UNIVERSITY PRESS
The Edinburgh Building, Cambridge CB2 2RU, UK   www.cup.cam.ac.uk
40 West 20th Street, New York, NY 10011–4211, USA   www.cup.org
10 Stamford Road, Oakleigh, Melbourne 3166, Australia
Ruiz de Alarcón 13, 28014 Madrid, Spain

First edition published by Chapman and Hall 1996
Second edition published 2000

Printed in the United Kingdom at the University Press, Cambridge

Typeface Palatino 10/12.5pt. *System* QuarkXPress™ [SE]

A catalogue record for this book is available from the British Library

Library of Congress Cataloguing in Publication data

Service, M. W.
Medical entomology for students / Mike W. Service. – 2nd ed.
    p.   cm.
Includes bibliography references and index.
ISBN 0 521 66659 7
1. Insects as carriers of disease.   I. Title.
RA639.5.S47   2000
614.4'32–dc21   99-16231   CIP

ISBN 0 521 66659 7 paperback

To Wednesday once again

Contents

Preface to second edition

This new edition remains largely unchanged in respect of area covered from the first edition, and I have generally kept to the same style and format. As stated in the first edition this is a student textbook on medical entomology; those interested more in reference books should consult the publications listed at the end of this Preface. I have revised the text where necessary and rewritten many of the sections on control as these can quickly become outdated. A number of new figures are included, and some previous ones have been redrawn or modified. A few of the older references under the headings 'Further reading' have been omitted and several new ones added. Finally I have added a Glossary, mainly entomological, that I hope will help students to understand better some of the entomological terms commonly used.

FURTHER READING

Beaty, B.J. and Marquardt, W.C. (eds.) (1996) *The Biology of Disease Vectors*. Colorado: University Press of Colorado.

Chavasse, D.C. and Yap, H.H. (eds.) (1997) *Chemical Methods for the Control of Vectors and Pests of Public Health Importance*. WHO/CTD/WHOPES/97.2. Geneva: World Health Organization.

Clark, G.G. (coordinator) (1994) Prevention of tropical diseases: status of new and emerging vector control strategies. Proceedings of a symposium on vector control. *American Journal of Tropical Medicine and Hygiene*, 50 (Suppl.), 159pp.

Kettle, D.S. (1995) *Medical and Veterinary Entomology*, 2nd edn. Wallingford: CAB International.

Lane, R.P. and Crosskey, R.W. (eds.) (1993) *Medical Insects and Arachnids*. London: Chapman & Hall.

Walker, A. (1994) *Arthropods of Humans and Domestic Animals: A Guide to Preliminary Identification*. London: Chapman & Hall.

World Health Organization (1997) *Vector Control: Methods for Use by Individuals and Communities*, prepared by J.A. Rozendaal. Geneva: World Health Organization.

Mike W. Service
Liverpool, February 1999

Preface to first edition

This is not intended as a reference book on medical entomology; those interested in such a book should consult *Medical Insects and Arachnids* (1993) edited by R.W. Lane and R.W. Crosskey (Chapman & Hall). The present book is aimed at students, whether they be physicians, nurses, health officials, community health workers or those studying for a masters' degree in parasitology or medical entomology. Its aim is to provide basic information on the recognition, biology and medical importance of arthropods and guidelines for their control. In a teaching book such as this it is always difficult to decide how much detail to include and what to omit; you cannot satisfy everyone. Nevertheless I have attempted to write a book to suit the needs of most students.

The reader should be selective. For example, I hope that most will find all, or most of, the information given in the chapters on fleas, lice, bedbugs, scabies mites and flies relevant to their needs, but I would expect readers to be more selective with some other chapters, such as those on mosquitoes. These insects are undoubtedly the most important arthropod vectors; nevertheless some students may think I have included too much detail on certain aspects for their needs, and if so they can largely disregard such bits. I have also tried to be selective and avoid giving too many references at the end of each chapter, but with some vectors this has not been easy.

Mike W. Service
Liverpool, May 1995

Acknowledgements

I would like to thank the following colleagues and friends, arranged in alphabetical order, for their valuable comments on various sections of the book: R.W. Ashford, J.P.T. Boorman, J.B. Davies, M.J.R. Hall, R.E. Harbach, P. Hillyard, A.M. Jordan, C.G. Moore, C.J. Schofield, R.D. Ward and M.R.G. Varma.

I am indebted to The Trustees of the Natural History Museum, London for permission to reproduce in a modified form Figs. 4.2, 4.4a, 17.1 (male tick) and 19.3.

1

Introduction to mosquitoes (Culicidae)

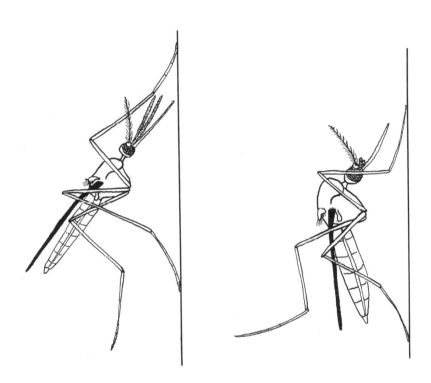

There are some 3200 species and subspecies of mosquitoes belonging to 37 genera, all contained in the family Culicidae. This family is divided into three subfamilies: Toxorhynchitinae, Anophelinae (anophelines) and Culicinae (culicines). Mosquitoes have a world-wide distribution; they occur throughout the tropical and temperate regions and extend their range northwards into the Arctic Circle. The only areas from which they are absent are Antarctica, and a few islands. They are found at elevations of 5500 m and down mines at depths of 1250 m below sea level.

The most important pest and vector species belong to the genera *Anopheles, Culex, Aedes, Psorophora, Haemagogus* and *Sabethes*.

Anopheles species, as well as transmitting malaria, are also vectors of filariasis (*Wuchereria bancrofti, Brugia malayi* and *Brugia timori*) and a few arboviruses. Certain *Culex* species transmit *Wuchereria bancrofti* and a variety of arboviruses. The genus *Aedes* contains important vectors of yellow fever, dengue, encephalitis viruses and many other arboviruses, and in a few restricted areas they are also vectors of *Wuchereria bancrofti* and *Brugia malayi*, whereas *Mansonia* species transmit *Brugia malayi* and sometimes *Wuchereria bancrofti* and a few arboviruses. *Haemagogus* and *Sabethes* mosquitoes are vectors of yellow fever and a few other arboviruses in Central and South America, while the genus *Psorophora* contains some troublesome pest species in North and South America.

Several mosquitoes in other genera have also been incriminated as vectors of various arboviruses. Moreover, many species, although not carriers of any disease, can nevertheless be troublesome because of the serious biting nuisances they cause.

1.1   EXTERNAL MORPHOLOGY OF MOSQUITOES

Mosquitoes possess only one pair of functional wings, the fore-wings. The hind-wings are represented by a pair of small, knob-like halteres. Mosquitoes are distinguished from other flies of a somewhat similar shape and size by: (i) the possession of a conspicuous forward-projecting proboscis; (ii) the presence of numerous appressed scales on the thorax, legs, abdomen and wing veins; and (iii) a fringe of scales along the posterior margin of the wings.

Mosquitoes are slender and relatively small insects, usually measuring about 3–6 mm in length. Some species, however, can be as small as 2 mm while others may be as long as 19 mm. The body is distinctly divided into a head, thorax and abdomen.

The head has a conspicuous pair of kidney-shaped compound eyes. Between the eyes arises a pair of filamentous and segmented antennae. In females the antennae have whorls of short hairs (that is pilose antennae), but in males, with a few exceptions in genera of no medical importance, the antennae have many long hairs giving them a feathery or plumose appearance. Mosquitoes can thus be conveniently sexed by examination of the antennae: individuals with feathery antennae are males, whereas those

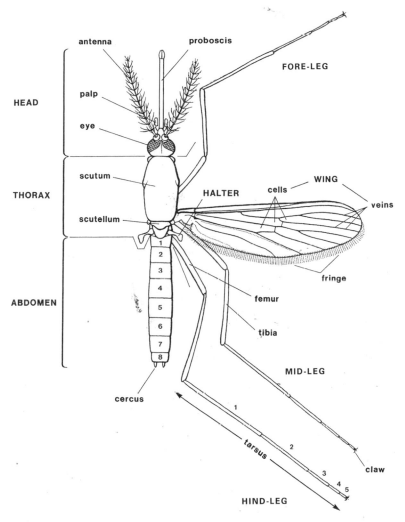

Figure 1.1 Diagrammatic representation of a female adult mosquito.

with only short and rather inconspicuous antennal hairs are females (Figs. 1.1, 1.13). Just below the antennae are a pair of palps which may be long or short and dilated or pointed at their tips, depending on the sex of the adults and whether adults are anophelines or culicines (Fig. 1.13). Arising between the palps is the single elongated proboscis, which contains the piercing mouthparts of the mosquito. In mosquitoes the proboscis characteristically projects forwards (Fig. 1.1).

The thorax is covered, dorsally and laterally, with scales which may be dull or shiny, white, brown, black or almost any colour. It is the arrangement of black and white, or coloured, scales on the dorsal surface of the thorax that gives many species (especially those of the genus *Aedes*) distinctive patterns (Fig. 3.3).

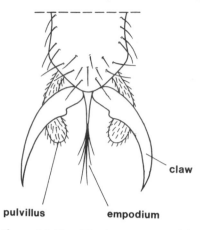

pulvillus   **empodium**

Figure 1.2 Tip of the last segment of the tarsus of a *Culex* mosquito showing claws, hair-like empodium and two large pulvilli.

The wings are long and relatively narrow and the number and arrangement of the wing veins is virtually the same for all mosquito species (Fig. 1.1). The veins are covered with scales which are usually brown, black, white or creamy yellow, but more brightly coloured scales may occasionally be present. The shape of the scales and the pattern they form differs considerably between both genera and species of mosquitoes. Scales also project as a fringe along the posterior border of the wings. In life the wings of resting mosquitoes are placed across each other over the abdomen in the fashion of a closed pair of scissors. The legs of the mosquito are long and slender and are covered with scales which are usually brown, black or white and may be arranged in patterns, often in the form of rings (Fig. 3.4). The tarsus terminates in, usually, a pair of toothed or simple claws. Some genera, such as *Culex*, have a pair of small fleshy pulvilli (Fig. 1.2) at the end of the tarsus.

The abdomen is composed of 10 segments but only the first seven or eight are visible. In mosquitoes of the subfamily Culicinae, the abdomen is usually covered dorsally and ventrally with mostly brown, blackish or whitish, scales. In the Anophelinae, however, the abdomen is almost, or entirely, devoid of these scales. The last abdominal segment of the female mosquito terminates in a pair of small finger-like cerci, whereas in the males a pair of prominent claspers, comprising part of the male external genitalia, are present.

In unfed mosquitoes the abdomen is thin and slender, but after females have bitten a suitable host and taken a blood-meal (only females bite) the abdomen becomes greatly distended and resembles a red oval balloon. When the abdomen is full of developing eggs it is also dilated, but is whitish and not red in appearance.

1.1.1   Mouthparts and salivary glands

The mouthparts are collectively known as the proboscis. In mosquitoes the proboscis is elongate and projects conspicuously forwards in both sexes – although males do not bite. The largest component of the mouthparts is the long and flexible gutter-shaped labium which terminates in a pair of small flap-like structures called the labella. In cross-section the labium is seen to almost encircle all the other components of the mouthparts (Fig. 1.3) and serves as a protective sheath. The individual components are held close together in life and only become partially separated during blood-feeding, or when they are teased apart for examination as illustrated in Fig. 1.4.

The uppermost structure, the labrum, is slender, pointed and grooved along its ventral surface. In between this 'upper roof' (labrum) and 'lower gutter' (labium) are five needle-like structures, namely, a lower pair of toothed maxillae, an upper pair of mandibles, which are more finely toothed, and finally a single untoothed, hollow stylet called the hypopharynx. When a female mosquito bites a host the labella, at the tip of the fleshy labium, are placed on the skin and the labium, which cannot pierce the skin, curves backwards. This allows the paired mandibles, paired maxillae, labrum and hypopharynx to penetrate the host's skin. Saliva from a pair of trilobed salivary glands, situated ventrally in the anterior part of the thorax, is pumped down the hypopharynx. Saliva, in at least some species, contains anticoagulants which prevent the blood from clotting and obstructing the mouthparts as it is sucked up into the space formed by the apposition of the labrum and other piercing mouthparts. Saliva also contains antihaemostatic enzymes that produce haematomas in the skin and facilitate the location and uptake of blood, and anaesthetic substances that help reduce the pain inflicted by the mosquito's bite, so reducing the host's defensive reactions.

Although male mosquitoes have a proboscis, the maxillae and mandibles are usually reduced in size or the mandibles are absent, so males cannot bite.

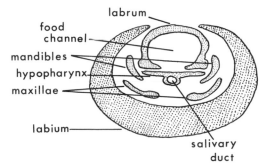

Figure 1.3 Diagram of a cross-section through the proboscis of a mosquito showing components of the mouthparts and the food channel.

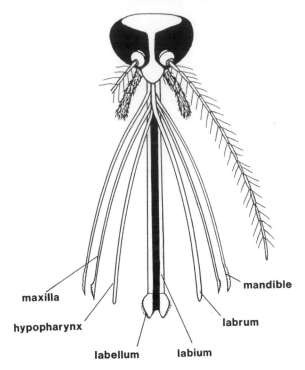

maxilla

hypopharynx

labellum   labium

mandible

labrum

Figure 1.4 Diagram of the head of a female culicine mosquito showing the components of the mouthparts spread out from the labium.

1.2   LIFE CYCLE

1.2.1   Blood-feeding and the gonotrophic cycle

Most mosquitoes mate shortly after emergence from the pupa. Spermatozoa, passed by the male into the spermatheca of the female, usually serve to fertilize all eggs laid during her lifetime; thus only one mating and insemination per female is required. With a few exceptions, a female mosquito must bite a host and take a blood-meal to obtain the necessary nutrients for the development of the eggs in the ovaries. This is the normal procedure and is referred to as anautogenous development. A few species, however, can develop the first batch of eggs without a blood-meal. This process is called autogenous development. The speed of digestion of the blood-meal depends on temperature and in most tropical species takes only 2–3 days, but in colder, temperate countries blood digestion may take as long as 7–14 days.

After a blood-meal the mosquito's abdomen is dilated and bright red in colour, but some hours later the abdomen becomes a much darker red. As the blood is digested and the white eggs in the ovaries enlarge, the abdomen becomes whitish posteriorly and dark reddish anteriorly. This

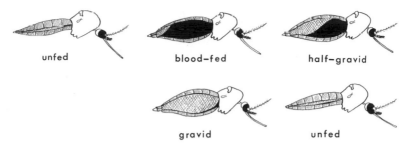

Figure 1.5 Diagrammatic representation of the gonotrophic cycle of a female mosquito. Each cycle starts with an unfed adult which passes through a blood-fed, half-gravid and gravid condition. After oviposition the female is again unfed and seeks another blood-meal.

condition represents a mid-point in blood digestion and ovarian development, and the mosquito is referred to as being half-gravid (Fig. 1.5). Eventually all blood is digested and the abdomen becomes dilated and whitish due to the formation of fully developed eggs (Fig. 1.5). The female is now said to be gravid and she searches for suitable larval habitats in which to lay her eggs. After oviposition the female mosquito takes another blood-meal and after 2–3 days (in the tropics) a further batch of eggs is matured. This process of blood-feeding and egg maturation, followed by oviposition, is repeated several times throughout the female's life and is referred to as the gonotrophic cycle.

Although male mosquitoes have a conspicuous proboscis, their mandibles and maxillae are insufficiently developed for piercing the skin and blood-feeding; instead they feed on the nectar of flowers, and other naturally occurring sugary secretions. Males are consequently unable to transmit any diseases. Sugar feeding is not, however, restricted to males; females may also feed on sugary substances to obtain energy for flight and dispersal, but only in a few species (the autogenous ones) is this type of food sufficient for egg development.

1.2.2   Oviposition and biology of the eggs

Depending on the species, female mosquitoes lay about 30–300 eggs at any one oviposition. Eggs are brown or blackish and 1 mm or less long. In many Culicinae they are elongate or approximately ovoid in shape, but eggs of *Mansonia* are drawn out into a terminal filament (Fig. 3.8). In the Anophelinae eggs are usually boat-shaped (Fig. 1.8). Many mosquitoes, such as species of *Anopheles* and *Culex*, lay their eggs directly on the water surface. In *Anopheles* the eggs are laid singly and float on the water, whereas those of *Culex* are laid vertically in several rows held together by surface tension to form an egg raft which floats on the water (Fig. 1.15). *Mansonia* species lay their eggs in a sticky mass that is glued to the underside of floating plants. None of the eggs of these mosquitoes can survive desiccation and consequently they die if they become dry. In the tropics eggs hatch within 2–3 days, but in cooler temperate countries they may not hatch until after 7–14 days, or longer.

Other mosquitoes, such as those belonging to the genera *Aedes*, *Psorophora* and *Haemagogus*, do not lay eggs on the water surface but deposit them just above the water line on damp substrates, such as mud and leaf litter, or on the inside walls of tree-holes and clay water-storage pots. Eggs of these genera can withstand desiccation, especially those of *Aedes* and *Psorophora* which can remain dry for months or even years but still remain viable and hatch when soaked in water. Because such eggs are laid above the water line of breeding places it may be many weeks or months before they become flooded with water, and thus have the opportunity to hatch. However, even when flooded, hatching may extend over long periods because the eggs hatch in instalments. Moreover, eggs of *Aedes* and *Psorophora* may require repeated immersions in water followed by short periods of desiccation before they will hatch. *Aedes* and *Psorophora* eggs may also enter a state of either quiescence, hatching when suitable conditions occur, or diapause and will not hatch until some specific stimulus terminates the state of diapause. Environmental stimuli such as changes in daylength and/or temperature often break diapause and cause eggs to hatch. In temperate regions many *Aedes* and *Psorphora* species overwinter as diapausing eggs.

1.2.3   Larval biology

Mosquito larvae can be distinguished from all other aquatic insects by being legless and having a bulbous thorax that is wider than both the head and abdomen. There are four active larval instars. All mosquito larvae require water in which to develop; no mosquito has larvae that can withstand desiccation although they may be able to survive short periods among, for example, wet mud.

Mosquito larvae have a well-developed head, bearing a pair of antennae and a pair of compound eyes. Prominent mouthbrushes are present in most species and serve to sweep water containing minute food particles into the mouth. The thorax is roundish in outline and has various simple and branched hairs which are usually long and conspicuous. The 10-segmented abdomen has nine visible segments, most of which have simple or branched hairs (Figs. 1.9, 1.16). The last segment, which differs in shape from the preceding eight segments, has two paired groups of long hairs forming the caudal setae, and a larger fan-like group comprising the ventral brush (Figs. 1.10, 1.16), and ends in two pairs of transparent, sausage-shaped anal papillae. Although often called gills they are not concerned with respiration but with osmoregulation.

Mosquito larvae, with the exception of *Mansonia* and *Coquillettidia* species (and a few other mosquito species), must come to the water surface to breathe. Atmospheric air is taken in through a pair of spiracles situated dorsally on the tenth abdominal segment. In the subfamilies Toxorhynchitinae and Culicinae these spiracles are situated at the end of a single dark-coloured and heavily sclerotized tube termed the siphon (Fig. 1.16). *Mansonia* and *Coquillettidia* larvae possess a specialized siphon that is more or less conical, pointed at the tip and supplied with prehensile hairs and serrated cutting structures (Fig. 3.9). These enable the siphon to be inserted into the roots or stems of aquatic plants and thus oxygen for larval respiration is obtained from the plants. In contrast larvae of the Anophelinae do not have a siphon (Figs. 1.10, 1.13).

Mosquito larvae feed on yeasts, bacteria, protozoans and numerous other plant and animal micro-organisms found in the water. Some, such as *Anopheles* species, are surface-feeders, whereas many others browse over the bottoms of habitats. A few mosquitoes are carnivorous or cannibalistic. There are four larval instars and in tropical countries larval development, that is the time from egg hatching to pupation, can be as short as 5–7 days, but many species require about 7–14 days. In temperate areas the larval period may last several weeks or months, and several species overwinter as larvae.

1.2.4   Larval habitats

Mosquito larval habitats vary from large and usually permanent collections of water, such as freshwater swamps, marshes, ricefields and borrow pits, to smaller collections of temporary water such as pools, puddles, water-filled car tracks, ditches, drains and gulleys. A variety of 'natural container-habitats' also provide breeding places, such as water-filled tree-holes, rock-pools, water-filled bamboo stumps, bromeliads, pitcher plants, leaf axils in banana, pineapple and other plants, water-filled split coconut husks and snail shells. Larvae also occur in wells and 'man made container-habitats', such as clay pots, water-storage jars, tin cans, discarded kitchen utensils and motor vehicle tyres. Some species prefer shaded larval habitats whereas others like sunlit habitats. Many species cannot survive in water polluted with organic debris whereas others can breed prolifically in water contaminated with excreta or rotting vegetation. A few mosquitoes breed almost exclusively in brackish or salt water, such as in salt-water marshes and mangrove swamps, and are consequently restricted to mostly coastal areas. Some species are less specific in their requirements and can tolerate a wide range of different types of breeding places.

Almost any collection of permanent or temporary water can constitute a mosquito larval habitat, but larvae are usually absent from large expanses of uninterrupted water such as lakes, especially if they have large numbers of fish and other predators which are likely to eat mosquito larvae. They are also usually absent from large rivers and fast-flowing waters, except that they may occur in marshy areas and isolated pools and puddles formed at the edges of flowing water.

1.2.5 Pupal biology

All mosquito pupae are aquatic and comma-shaped. The head and thorax are combined to form the cephalothorax, which has a pair of respiratory trumpets dorsally (Fig. 1.6). The abdomen is 10-segmented although only eight segments are visible. Each segment has numerous short hairs and the last segment terminates in a pair of oval and flattened structures termed paddles (Figs. 1.11, 1.18). Some of the developing structures of the adult mosquito can be seen through the integument of the cephalothorax, the most conspicuous features being a pair of dark compound eyes, folded wings, legs and the proboscis (Fig. 1.6).

Pupae do not feed but spend most of their time at the water surface taking in air through the respiratory trumpets. If disturbed they swim up and down in the water in a jerky fashion.

Pupae of *Mansonia* and *Coquillettidia* differ in that they have relatively long breathing trumpets, which are modified to enable them to pierce aquatic vegetation and obtain their oxygen in a similar fashion to the larvae (Fig. 3.9). As a consequence their pupae remain submerged and rarely come to the water surface.

In the tropics the pupal period in mosquitoes lasts only 2–3 days but in cooler temperate regions pupal development may be extended over 9–12

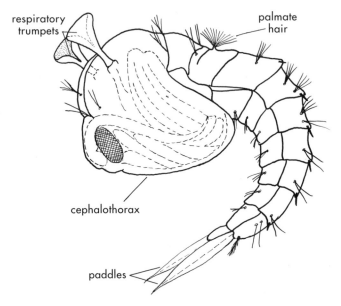

Figure 1.6 *Anopheles* pupa.

days, or longer. At the end of pupal life the skin on the dorsal surface of the cephalothorax splits, and the adult mosquito struggles out.

1.2.6 Adult biology and behaviour

As already mentioned (p. 6), females of most species of mosquito require a blood-meal before the eggs can develop, and this is taken either before or more usually after mating. Many species bite humans to obtain their blood-meals and a few feed on humans in preference to other animals. However, others prefer feeding on non-human hosts and many species never bite people. Species that usually feed on humans are said to be anthropophagic in their feeding habits, whereas those that feed mainly on other animals are called zoophagic. Mosquitoes that feed on birds are sometimes called ornithophagic instead of zoophagic. Females are attracted to hosts by various stimuli emanating from them, such as body odours, carbon dioxide and heat. Vision usually plays only a minor role in host orientation but some species are attracted by the silhouette or movement of their hosts. Some species feed more or less indiscriminately at any time of the day or night; others are mainly diurnal or nocturnal in their biting habits.

A few species of mosquitoes frequently enter houses to feed and are said to be endophagic in their feeding habits, whereas those that bite their hosts outside houses are called exophagic. After having bitten humans, or some other hosts, either inside or outside houses, mosquitoes seek resting places in which to shelter during digestion of their blood-meals. Some species rest inside houses during the time required for blood digestion and maturation of the ovaries and are called endophilic. In contrast mosquitoes that rest outdoors are termed exophilic. Female adults of *Aedes aegypti* (a vector of yellow fever), for example, are usually anthropophagic, exophagic and exophilic, whereas adults of *Anopheles gambiae* (African malaria vector) are mainly anthropophagic, endophagic and endophilic. Few mosquitoes, however, are entirely anthropophagic or zoophagic, endophagic or exophagic, endophilic or exophilic. Instead, most show various degrees of these behavioural patterns; in other words all these terms are relative. The feeding behaviour of a species may also change; for example in certain areas and at certain seasons a species may bite people predominantly (anthropophagic) inside houses (endophagic) and remain in houses afterwards (endophilic), whereas at other times, especially if there are few people but many animals in the area, the species may become predominantly zoophagic, and also exophagic and exophilic. Some species are less adaptable in their feeding behaviour and will never rest inside houses or enter them to feed on people.

The biting behaviour of female mosquitoes may be very important in the epidemiology of disease transmission. Mosquitoes that feed on people predominantly out of doors and late at night, will not bite many young children because they will be indoors and asleep at this time. Thus, young children will be less likely to be infected with any diseases that these mosquitoes might transmit. During hot and dry periods of the year substantial numbers of people may sleep out of doors and as a consequence be bitten more frequently by exophagic mosquitoes. Some mosquitoes bite predominantly within forests or wooded areas. Consequently, humans will only get bitten when they visit these places. Clearly the behaviour of both people and mosquitoes may be relevant in diseases transmission.

The resting behaviour of adult mosquitoes may be an important consideration in planning control measures. In several malaria control campaigns the interior surfaces of houses, such as walls and ceilings, are sprayed with residual insecticides, such as DDT, to kill adult mosquitoes resting on them. This approach will, of course, only be effective in controlling malaria if the mosquito vectors are endophilic.

Most mosquitoes probably disperse only a few hundred metres from their emergence sites, and in control programmes and epidemiological studies it is usually safe to say that mosquitoes will not fly further than 2 km. There are, however, records of mosquitoes being found up to 100 km or more from their breeding places, but such dispersal is nearly always wind assisted. Mosquitoes may get transported long distances in aeroplanes, and sometimes this causes disease outbreaks, such as 'airport malaria'.

In tropical countries adult female mosquitoes probably live *on average* 1–2 weeks, whereas in temperate countries adult longevity is likely to be 3–4 weeks, but some adults must live longer otherwise they would not be vectors of diseases such as malaria. Species that hibernate or aestivate live much longer; for example in Europe some fertilized females of *Culex pipiens* survive in hibernation from August until May. Male adults usually have a shorter life-span than females.

1.3   CLASSIFICATION OF MOSQUITOES

1.3.1   Subfamily Toxorhynchitinae

Subfamily Toxorhynchitinae comprises a single genus, *Toxorhynchites*, which contains 73 species which are mainly tropical, although a few species occur in North America, eastern Russia and Japan.

Adults are large (19 mm long, 24 mm wing spread) and colourful, being metallic bluish or greenish with black, white or red tufts of hair-like scales projecting from the posterior abdominal segments. Adults are easily recognized by the possession of a proboscis that is curved backwards in both sexes (Fig. 1.7) and is incapable of piercing the skin. Consequently, since neither sex can bite, they are of no medical importance. Their larvae are also large (12–18 mm long), often dark reddish and, like those of the Culicinae, have a siphon. They are predacious on larvae of other mosquitoes and on their own kind. They have occasionally been introduced into areas in the hope that their voracious larvae will help reduce the numbers

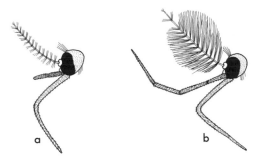

Figure 1.7 Heads of *Toxorhynchites* adults: (*a*) female; (*b*) male.

of pest mosquitoes. Larvae are found mainly in container-habitats, such as tree-holes and bamboo stumps, tin cans and water-storage pots.

1.3.2 Subfamily Anophelinae

Of the three genera included in the subfamily Anophelinae only the genus *Anopheles* (about 430 species), which contains important malaria vectors, is of medical importance. The following characters serve to separate *Anopheles* mosquitoes from those in any of the other mosquito genera.

Anopheline eggs

Eggs are laid singly on the water surface. In most species they are typically boat-shaped and, laterally, have a pair of floats (Fig. 1.8). Anopheline eggs are unable to withstand desiccation.

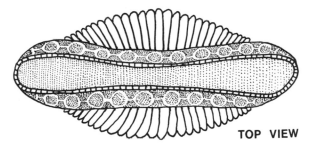

TOP VIEW

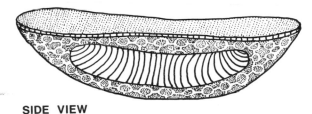

SIDE VIEW

Figure 1.8 *Anopheles* eggs.

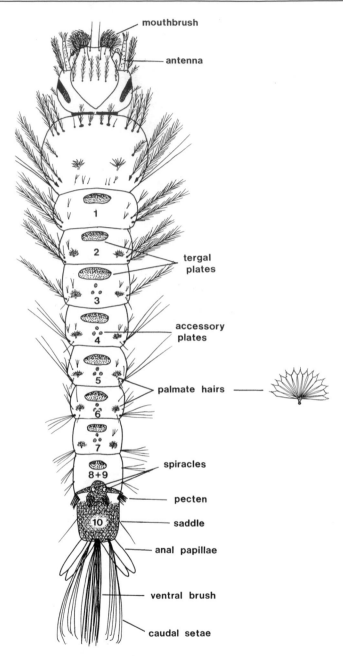

Figure 1.9 *Anopheles* larva, dorsal view, showing diagnostic abdominal tergal plates and palmate hairs.

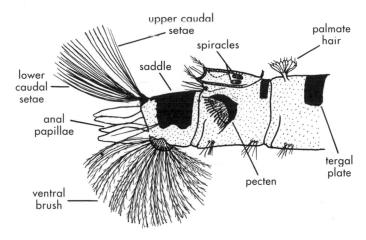

Figure 1.10 Abdominal terminal segments of an *Anopheles* larva in lateral view.

Anopheline larvae

Larvae lack a siphon and lie parallel to the water surface, not subtended at an angle as are the culicines. They are surface-feeders and spend most of their time at the water surface. Examination under a microscope shows that the abdomen has small, brown, sclerotized plates, called tergal plates, on the dorsal surface of abdominal tergites 1–8. In addition most or all of these segments have a pair of well-developed palmate hairs, sometimes called float hairs (Figs. 1.9, 1.10). These abdominal palmate hairs and the single pair on the thorax come into contact with the water surface and aid in keeping the larvae parallel to the surface. All these structures identify larvae as belonging to the genus *Anopheles*.

Anopheline pupae

The respiratory trumpets of anopheline pupae are short and broad distally, thus appearing conical (Figs. 1.6, 1.11*a*), whereas in most culicines the trumpets are narrower and more cylindrical. The most reliable character for identifying anopheline pupae is the presence of short, peg-like spines situated laterally near the distal margins of abdominal segments 2–7 or 3–7 (Fig. 1.11*b*); in the culicines there are no such spines.

Anopheline adults

Adult *Anopheles* usually rest with their bodies at an angle to the surface, that is with the proboscis and abdomen in a straight line (Fig. 1.13). In some species they rest at almost right angles to the surface, whereas in others such as the Indian mosquito *Anopheles culicifacies*, the angle is much smaller. This is a very useful characteristic allowing adults resting in

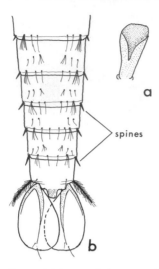

Figure 1.11 *Anopheles* pupa. (*a*) Short broad respiratory trumpet; (*b*) characteristic spines on abdomen.

houses and elsewhere to be readily identified as *Anopheles*. Most, but not all, *Anopheles* mosquitoes have the dark (usually blackish) and pale (usually whitish or creamy white) scales on the wing veins arranged in 'blocks' or specific areas (Fig. 1.12) forming a distinctive spotted pattern which differs according to species. A few species, however, such as the European *An. claviger*, have the veins covered more or less uniformly with dark (often brown) scales.

The most reliable way to distinguish between adult *Anopheles* and Culicinae is by examination of their heads. The first procedure is to determine the sex of the adults: female mosquitoes have non-plumose antennae whereas males have plumose antennae. If the adults are females and also *Anopheles* then the palps will be about as long as the proboscis and usually lie closely alongside it (Fig. 1.13). The palps are usually blackish with broad or narrow rings of pale scales, especially on the apical half. In male *Anopheles* the palps are also about as long as the proboscis but are distinctly swollen at the ends and are said to be clubbed (Fig. 1.13); they may also have rings of pale scales apically.

Other minor differences between *Anopheles* and the Culicinae are that (i) in both sexes of *Anopheles* the scutellum is rounded posteriorly and has setae along the entire edge; (ii) only one spermatheca is present in the females; and (iii) in both sexes the middle lobe of the salivary glands is considerably shorter than the two outer lobes (Fig. 1.14).

The principal characters for separating the various stages in the life cycles of anopheline and culicine mosquitoes are given in Table 1.1.

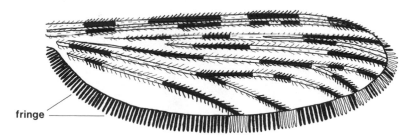

fringe

Figure 1.12 *Anopheles* wing showing dark and pale scales on veins arranged in 'blocks'.

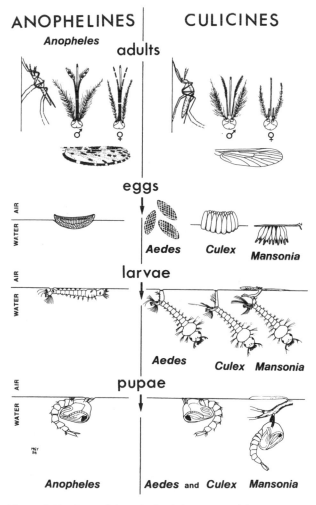

Figure 1.13 Chart of the principal characters of the various stages in the life cycle that distinguish anopheline from culicine mosquitoes.

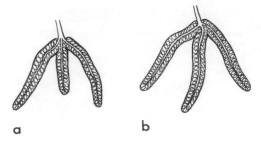

a   b

Figure 1.14 Salivary glands of adult mosquitoes. (*a*) *Anopheles*; (*b*) culicine.

Table 1.1.   Principal characters distinguishing anopheline and culicine mosquitoes

Stage	Anophelinae	Culicinae
Eggs	Laid singly, possess floats	Laid singly or in egg rafts or masses. Never possess floats
Larvae	Never have a siphon. Lie parallel to water surface. Have abdominal palmate hairs and tergal plates	All larvae have a short or long siphon. Subtend an angle from the water surface. No palmate hairs or tergal plates
Pupae	Breathing trumpets short and broad apically. Short peg-like abdominal spines on segments 2–7 or 3–7	Breathing trumpets short or long, opening not broad. No spines on abdominal segments 2–7
Adults (both sexes)	Rest at an angle to any surface. In most species dark and pale scales on wing veins arranged in distinct 'blocks'	Rest with the bodies more or less parallel to the surface. Scales on wing veins not arranged in 'blocks'; scales frequently all brown or blackish, or a mixture of pale and dark scales scattered on veins
Adult females (non-plumose antennae)	Palps about as long as proboscis	Palps much shorter than proboscis
Adult males (plumose antennae)	Palps about as long as proboscis and swollen at ends	Palps about as long as proboscis but never swollen at ends; may be hairy distally

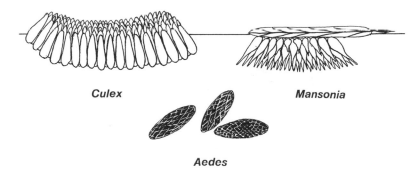

Culex   *Mansonia*

Aedes

Figure 1.15 Mosquito eggs. *Culex* egg raft floating on the water surface, *Mansonia* eggs glued to the undersurface of floating vegetation, and individual *Aedes* eggs, which are deposited on damp surfaces.

1.3.3 Subfamily Culicinae

There are some 2700 species in the large subfamily Culicinae, belonging to 34 genera. The most important medically are the genera *Aedes*, *Culex*, *Mansonia*, *Haemagogus*, *Sabethes* and *Psorophora*. The following characters serve to separate the Culicinae from *Anopheles* mosquitoes. Methods for distinguishing between the more important genera within the Culicinae are given in Chapter 3.

Culicine eggs

Culicine eggs never have floats. They are laid either singly (e.g. *Aedes*) or in the form of egg rafts that float on the water surface (e.g. *Culex*), or are deposited as sticky masses glued to the underside of floating vegetation (e.g. *Mansonia*) (Fig. 1.15).

Culicine larvae

All culicine larvae possess a siphon (Fig. 1.16), which may be long or short. They hang upside down at an angle from the water surface when they are getting air (Fig. 1.13), except for *Mansonia* and *Coquillettidia* larvae which insert their specialized siphons into aquatic plants and remain submerged (Fig. 3.9). There are no abdominal palmate hairs or tergal plates on culicine larvae.

Culicine pupae

The length of the respiratory trumpets in culicine pupae is variable, but the trumpets are generally longer, more cylindrical and their openings narrower (Fig. 1.17*b*) than in *Anopheles*. Abdominal segments 2–7 lack peg-like spines although they have numerous setae (Fig. 1.18).

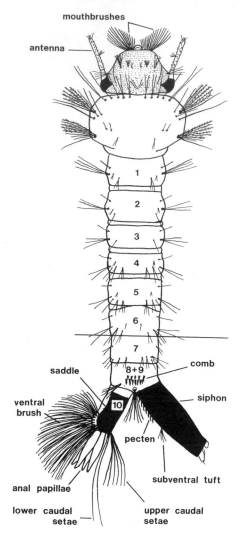

Figure 1.16 Culicine larva, dorsal view but abdominal segments 7–9 turned laterally to display important characters.

Culicine adults

In life, culicine adults rest with the thorax and abdomen more or less parallel to the surface (Fig. 1.13). The wing veins are commonly covered with scales of a uniform brown or black colour. Sometimes, however, there are contrasting dark and pale scales but they are not arranged in distinctive areas or 'blocks', as found in many *Anopheles* adults.

The most reliable method for identifying the Culicinae is to examine their heads. In females (which have non-plumose antennae), the palps

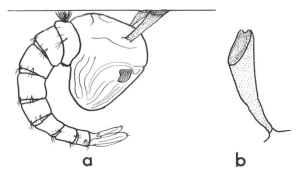

Figure 1.17 Culicine pupa. (a) Pupal position at the water surface; (b) elongated and relatively narrow respiratory trumpets.

are shorter than the proboscis. In males (which have plumose antennae), the palps are about as long as the proboscis but are not swollen distally and hence do not appear clubbed (Fig. 1.13). However, they may be turned upwards distally, and in many species they are covered with long hairs so that, superficially, they can appear to be somewhat swollen apically; more careful inspection shows that the palps in male culicines are not clubbed.

Other minor differences that separate the Culicinae from *Anopheles* are that in culicines the scutellum is trilobed and the scutellar setae are restricted to these lobes, whereas in anophelines the scutum is evenly rounded posteriorly and has setae along its entire edge. In culicines there are two or three spermathecae in the females but in anophelines just one, and in culicines the middle lobe of the salivary glands is about as long as the other two, whereas in anophelines it is shorter (Fig. 1.14).

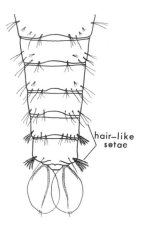

Figure 1.18 Abdomen of a culicine pupa showing hair-like setae (note absence of spines).

1.4   MEDICAL IMPORTANCE

Although in many temperate countries mosquitoes may be of little or no importance as disease vectors they can, nevertheless, cause considerable annoyance because of their bites. The greatest numbers of mosquitoes are found in the northern areas of the temperate regions, especially near or within the Arctic Circle, where the numbers biting can be so great at certain times of the year as to make almost any outdoor activity impossible. Because of their elongated mouthparts female mosquitoes have little difficulty in biting through clothing such as socks, shirts, trousers and woollen garments, but clothing with a much closer weave of material may prevent this.

Mosquitoes are important as vectors of malaria, various forms of filariasis and numerous arboviruses, the best known being dengue and yellow fever. Their role in the transmission of these diseases is discussed in Chapters 2 and 3.

1.5   MOSQUITO CONTROL

Greater efforts have been made to control mosquitoes than any other biting insects and a vast literature has accumulated on control operations.

Control measures which are directed against specific vectors, such as *Aedes aegypti*, *Culex quinquefasciatus* and the malaria vectors, are described in more detail in Chapters 2 and 3; only the broader principles of control are outlined here.

Control measures can be directed at either the immature aquatic stages or the adults, or at both stages simultaneously.

1.5.1   Control directed at the immature stages

Biological control

Although often termed naturalistic control there is little, if anything, natural about the process of biological control. Either the incidence of predators, parasites or pathogens in habitats must be greatly increased to obtain worthwhile control, or they have to be introduced into habitats from which they were originally absent; such environmental manipulations are not natural. Biological control of mosquitoes was very popular during the early part of this century, but with the development and availability of chemicals such as the organochlorines and organophosphates it was replaced by insecticidal control. However, because of problems with insecticide resistance and greater awareness of environmental contamination there has been renewed interest in biological (biocontrol) methods. They are, however, usually more difficult to implement and maintain than insecticidal methods. Moreover, if predators are used it is unlikely that they will prey exclusively on mosquito larvae and pupae but also eat harmless or beneficial insects. Finally, biological control does not lead to rapid control; it takes some days, or more likely weeks, before pest or vector mosquito populations are reduced in size.

Predators Larvivorous fish are the most widely used biological control agents, the most common being the top minnow or mosquitofish, which exists as two subspecies, namely *Gambusia affinis affinis* and *G. affinis holbrooki*. These are warm-water fish originally native to the southern USA but they have been introduced to over 60 countries worldwide, including the Pacific islands, Europe, the Middle East, India, South-east Asia and Africa, in attempts to control mosquito larvae. They are cannibalistic and aggressive fish which have sometimes destroyed native fish; consequently they should not be introduced into new areas. Another commonly used fish is the South African guppy, *Poecilia reticulata*, which is not so voracious as *G. affinis* but can tolerate organic pollution better and is more heat tolerant. There are numerous other types of fish that have been used to eat mosquito larvae, such as carp (e.g. *Cyprinus carpio*) in Chinese ricefields, an edible catfish (*Clarias fuscus*) in water-storage tanks in Myanmar to control *Aedes aegypti*, and *Oreochromis* (=*Tilapia*) species in Africa and *Aplocheilus* species in Europe and Asia.

Some predatory fish, such as *Aphanius dispar* and *Fundulus* species, breed in saline waters and can therefore be introduced into salt-water habitats. Fish are unsuitable for the control of mosquitoes which breed in small containers, pools and puddles that rapidly dry out. However, some fish, such as species of *Nothobranchius* and *Cynolebias*, which are the so-called instant or annual fish, have drought-resistant eggs and these are more suitable for introducing into small temporary habitats that repeatedly dry out.

Although fish have sometimes greatly reduced the numbers of larvae in certain habitats, such as borrow pits, ponds and ricefields, they have rarely proved effective in reducing the size of mosquito populations over large areas. Nor is there usually much convincing evidence that they have significantly decreased the incidence of mosquito-borne diseases, although in a few mainly rather arid areas in Iran, Afghanistan, Somalia, Ethiopia, Greece, former USSR and China, they have been credited with reducing malaria transmission.

Other predators of mosquito larvae and pupae include tadpoles of frogs and toads and various aquatic insect larvae, but these have rarely proved effective as control agents. A few mosquitoes have predacious larvae, for example *Toxorhynchites* species. These have been introduced into container-habitats in certain areas (e.g. Fiji, Samoa and Hawaii) to control larvae of other container-breeding mosquitoes but results have not been very encouraging.

Pathogens and parasites There are numerous pathogens, such as viruses (e.g. iridescent and cytoplasmic polyhedrosis viruses), bacteria (e.g. *Bacillus thuringiensis* subsp. *israelensis* (=*B. thuringiensis* serotype H-14) and *B. sphaericus*), protozoans (e.g. *Nosema vavraia, Thelohania*) and fungi (e.g. *Coelomomyces, Lagenidium, Culicinomyces*) that cause larval mortality. There are also several parasitic nematodes that kill mosquito larvae and the most promising for control purposes is *Romanomermis culicivorax*, which has been commercially mass-produced. These parasites and pathogens appear harmless to humans.

Although there is considerable interest in biological agents, with the exception of *B. thuringiensis* subsp. *israelensis (B.t.i.)* they cannot yet be routinely advocated for control, especially in tropical countries. *Bacillus thuringiensis* subsp. *israelensis* is undoubtedly the most useful pathogen as it can be easily mass-produced, is toxicologically safe to humans and wildlife, and is more or less specific in killing mosquito larvae (but it can also kill simuliids: see Chapter 4). It is formulated as a powder that is sprayed on the water surface and kills mosquito larvae when it is ingested. Mortality is caused by a delta-endotoxin. Genetic engineering using both non-recombinant and recombinant techniques is improving the larvicidal activity of the bacillus and in addition has transferred the genes responsible for production of the poisonous endotoxin into other bacteria, including *B. sphaericus*. Because the bacteria used are inactive there is no multiplication of the pathogen and consequently there have to be repeated applications, as with most larvicides. It is, therefore, more of a microbial insecticide than a true biological (living) agent that recycles in the environment.

Genetic control

Although genetic control methods are directed against the adults rather than the immature stages it is convenient to discuss this strategy here because it is really a form of biological control.

The most common method of genetic control involves the release in the field of sterile male mosquitoes that have been reared in the laboratory. Sterilization can be achieved by ionizing radiation, crossing closely related species to produce infertile hybrid males, or more usually by introducing chemosterilants into breeding trays containing the mosquito larvae which makes the emerging adults (both sexes) sterile. The aim is to introduce large numbers of healthy, but sterile, insectary-reared males into field populations that will compete with natural fertile males for female mates, so resulting in large numbers of infertile inseminations. The eggs layed by such females will be sterile and fail to hatch, thus causing a reduction in and, it is hoped, elimination of, the vector. In El Salvador the release in the 1970s of some 4.36 million chemosterilized male *Anopheles albimanus* (an important malaria vector that had developed resistance to most insecticides) over 4.5 months in an isolated coastal region of about 15 km^2 caused a more than 97% reduction in the biting population. However, because enormous numbers of mosquitoes had to be reared to obtain control over a very limited area, the later expanded trial failed. There is presently little interest in this technique for controlling mosquitoes.

Other methods of genetic control involve the use of cytoplasmic incompatibility, translocations, introduction of lethal genes, or genes that make mosquitoes refractory as a vector, and meiotic drive leading to the production of excessive numbers of males. Scientists skilled in genetic engineering have been trying to introduce undesirable genes into mosquitoes that will spread rapidly among natural populations and cause reductions in population size. More recently there has been interest in trying to replace natural vector populations with laboratory-reared and released mosquitoes of the same species that have been genetically altered to make them incapable of transmitting parasites. However, none of these genetic approaches is simple and they will be more difficult to implement than conventional insecticidal methods.

Physical control

Filling in, drainage and source reduction This is sometimes referred to as mechanical or environmental control. A simple form consists of filling in, and thus completely eradicating, breeding places. Larval habitats ranging in size from water-filled tree-holes to ponds and small marshes can be filled in with rubble, earth or sand. Filling in tree-holes can be difficult because many can be at great heights and difficult to locate, or there could be just too many for this method to be practical. Various container-habitats such as abandoned tin cans, metal drums, disused water-storage pots and old tyres can be removed; this approach is often referred to as source reduction. Mosquito breeding in water-storage pots that are in use can be reduced by covering up their openings, but this simple practice is not popular with the owners and it soon becomes neglected. The introduction of a reliable piped water supply should help reduce people's dependence on water-storage containers and thereby reduce breeding of mosquitoes such as *Aedes aegypti*, but in many areas with piped water people continue to store water in containers. In the Indian subcontinent water tanks are commonly sited on rooftops and are important breeding places of the vector (*Anopheles stephensi*) of urban malaria. Fitting these with mosquito screening would prevent breeding, but such covers usually become torn or are removed.

Some mosquitoes, such as *Culex quinquefasciatus*, breed in damaged septic tanks and soak-away pits, but this can be prevented if the tanks and pits are repaired so that egg-laying females cannot gain access. This mosquito also commonly breeds in pit latrines, but this can be stopped if small (2–3 mm) expanded polystyrene beads are tipped into latrines to form a floating layer about 1–2 cm thick. This prevents mosquitoes laying their eggs on the water surface.

Larval breeding places such as ponds, borrow pits, fresh- and salt-water marshes can be drained. An important advantage of filling in, draining or removing larval habitats is that such measures can lead to permanent control, but this approach is not always feasible. It is impossible, for example, to fill in all the scattered, small and temporary collections of water such as pools, vehicle tracks and puddles which may appear during the rainy season. Larger and more permanent habitats such as swamps may prove too costly to drain. Moreover, the local people may, understandably, not want certain breeding places filled in if the water is needed for domestic purposes or the sites used as watering points for livestock; such habitats may be an essential part of their life. The feasibility of eradicating breeding places must be assessed individually in each area.

Environmental manipulation If it is not feasible to eliminate mosquito breeding places it may be possible to alter them to make them unsuitable as larval habitats. For example, some mosquitoes breed in isolated pools and small marshy areas that form at the edges of ditches and streams with winding courses. Realigning these water courses to increase water flow and prevent the build-up of static pockets of water can greatly reduce mosquito breeding. The periodic opening of sluice gates has been practised in India and Malaysia to flush out larvae from small isolated pools of water. Other environmental modifications include the removal of overhanging vegetation to reduce breeding by shade-loving mosquitoes; conversely planting vegetation along reservoirs and streams may eliminate sun-loving species. Intermittent flooding of ricefields to allow drying out every 3–5 days can substantially reduce populations of several important vectors. Removal of rooted or floating vegetation will prevent breeding of *Mansonia* species because they require plants to obtain their oxygen requirements.

Instead of draining marshy areas they can be excavated to form areas of relatively deep permanent water with well-defined vertical banks. This process is called impoundment. It completely alters the habitat, making it unsuitable for many mosquitoes, especially *Aedes* and *Psorophora* species which lay their eggs on wet muddy edges of pools that are scattered over extensive marshy areas. Both small and large, fresh- and salt-water marshy areas can be converted into impounded waters. Sometimes such impounded waters are stocked with fish and ducks, and these animals may also help reduce mosquito breeding.

There is the danger, however, that whenever a larval habitat has been modified to reduce breeding of certain mosquitoes, but not eliminated, the new conditions created may now favour breeding of other mosquito species that were previously either absent or uncommon.

Chemical control

Most control directed against mosquitoes, except malaria vectors, consists of applying larvicides.

Oils One of the oldest control methods consists of spraying mineral oils onto the water surface of breeding places to kill mosquito larvae by poisoning and suffocation. Diesel oil and kerosene (paraffin) applied at the rate of 140–190 litres/ha have been used for years to control mosquito breeding. The addition of detergents, such as 0.5% octoxinol, or 1–2.5% vegetable oils (e.g. castor oil, coconut oil) increases the spreading power of the oils allowing application rates to be reduced to 18–50 litres/ha. Specially prepared commercial oils (e.g. Malariol, Flit MLO) have improved spreading properties so that only 10–20 litres/ha need be used. The addition of organophosphate or carbamate insecticides to high-spreading oils greatly enhances their effectiveness. Other useful larvicides include monolayer films of lecithins and various non-petroleum oils (e.g. AROSURF-MSF) which interfere with the properties of the air–water interface and cause larvae, pupae and even emerging or ovipositing adults to drown.

Oils have to be sprayed on breeding places about every 7–10 days in most tropical countries to ensure that larvae hatching from eggs are killed before they pupate and give rise to adults. Less frequent applications can be made in cooler temperate areas because the speed of development of the immature stages is much slower.

Paris Green Another old control method is the application of a fine dust of Paris Green (copper acetoarsenite) to breeding places. When ingested by surface-feeding mosquito larvae, such as *Anopheles* species, it acts as a stomach poison. Paris Green can also be formulated as pellets or granules that either float on the water surface or sink. The latter formulation can be effective against culicine species whose larvae are mainly bottom-feeders. Although mosquito larvae have not developed resistance to Paris Green this insecticide has largely been replaced by the organophosphate and carbamate insecticides.

As with oils repeated applications are usually needed.

Insecticides With the arrival of residual insecticides such as DDT in the mid-1940s the use of oils and Paris Green was more or less abandoned in favour of spraying larval habitats with these more modern chemicals.

However, because of their persistence in the environment and accumulation in food-chains DDT and other organochlorine insecticides should not be used as larvicides. In their place, less persistent and biodegradable insecticides should be used, such as the organophosphates and carbamates.

Recommended chemicals for larviciding include organophosphates such as malathion, pirimiphos methyl (Actellic), fenitrothion (Sumithion) and temephos (Abate), and carbamates such as propoxur (Baygon). Pyrethroids such as permethrin and deltamethrin can be used as larvicides but because they tend to kill other aquatic insects, crustaceans and even fish they should be used with caution and only in special cases. In organically polluted waters insecticides are less effective and either higher dosage rates must be used or the more effective organophosphates such as fenthion (Baytex) or chlorpyrifos (Dursban) applied. Chlorpyrifos is more toxic to mosquito larvae than most other insecticides, but it causes higher mortalities among fish and other aquatic organisms, and is consequently not so widely used.

All the above insecticides usually have to be sprayed on breeding places in tropical areas every 10–14 days, and more frequently on highly polluted waters.

Temephos (Abate) has very low mammalian toxicity and briquettes, granules or microencapsulated formulations which slowly release the insecticide over days or even weeks can be placed in containers holding potable water to control *Aedes aegypti*. However, in Yangon (Rangoon), Myanmar, people have refused to have their water pots treated with temephos, because they consider any insecticide in drinking water as environmental contamination. This attitude is likely to spread.

Mansonia larvae can be killed by spraying herbicides, such as diquat and 2,4-D, to kill the aquatic vegetation on which they rely to obtain their oxygen.

Larvicides are usually applied as emulsions or oil solutions, but granules (0.25–0.6 mm) or pellets (0.6–2 mm) are better for penetrating dense growths of aquatic vegetation. Insecticides formulated as slow-release granules or pellets can be scattered over marshy areas when they are relatively dry, then when they become flooded larvae hatching from drought-resistant aedine eggs, such as those of *Aedes* and *Psorophora* species, are killed as the granules release their toxicants into the water. Larvicides are usually delivered from knapsack-type sprayers carried on the backs of operators, but they are sometimes dispersed from spraying machines mounted in the backs of landrovers or trucks. Large or inaccessible areas may require aerial spraying from helicopters or small fixed-wing aircraft.

Insect growth regulators (IGRs) such as methoprene (Altosid) and pyriproxyfen which arrest larval development, or diflubenzuron (Dimilin) which inhibits chitin formation of the immature stages, are also used. These chemicals have the benefit of being environmentally friendly because they are more or less specific in killing mosquitoes and possess extremely low toxicity to humans. Methoprene is considered by the WHO to be safe for use in drinking water. However, the relatively high cost of IGRs limits their use.

Integrated control

It has become fashionable to advocate integrated control, which usually means combining biological and insecticidal methods: for example, the introduction of predacious fish to breeding places which are also sprayed with insecticides that have minimum effect on the fish. However, it is better to regard integrated control as any approach that takes into consideration more than one method, whether these are directed at only the larvae or adults, or both.

1.5.2 Control directed at adults

Personal protection

Much can be done to reduce the likelihood of being bitten by mosquitoes. Houses, hospitals and other buildings can have windows, doors and ventilators covered with mosquito screening, made of either strong plastic or non-corrosive metal. It is essential that screening is kept in good repair. Screens with 6–8 meshes/cm will exclude most mosquitoes. Finer-mesh screening will keep out smaller biting flies, some of which may be vectors, but will appreciably reduce ventilation and light. If houses are unscreened, or if screening is defective, mosquito nets with 9–10 meshes/cm can be used to protect against night-biting mosquitoes. Nets should be tucked in under mattresses or bedding, never allowed to drape loosely over beds. Torn nets are useless unless they have been impregnated with pyrethroid insecticides such as permethrin (see p. 49). Nets should be placed over beds before sunset. The main disadvantage of nets is that they reduce ventilation.

Small spray-guns (e.g. flit-guns) filled with pyrethrum or permethrin dissolved in kerosene or in an equal mix of kerosene and white spirit, or alternatively pressurized aerosol canisters containing pyrethroids, are commonly used to spray bedrooms early in the evening. Mosquito coils impregnated with pyrethroid insecticides, especially fast-acting ones like allethrin, which when ignited smoulder for 6–10 hours to produce an insecticidal smoke are commonly used in tropical countries. A more sophisticated method is to place small insecticide-impregnated tablets (vapourizing mats) on a mains-operated electric mini-heater.

Suitable insect repellents are diethyltoluamide (deet), *N,N*-diethylphenylacetamide (depa) and dimethylphthalate (dimp), which under optimal conditions can provide protection for 6–10 hours. Citronella oil can give good protection against mosquitoes, but only for about an hour; however, a repellent derived from lemon eucalyptus is said to be better – in fact almost as good as deet. Sweating and rubbing usually reduce the period of effectiveness. Repellents are applied to the hands, arms, neck and face (taking care to avoid the eyes), and the ankles and legs, irrespective of whether socks or long trousers are worn. More recently locally made repellent soap (includes 20% deet and 0.5% permethrin) has been shown to be effective, although the greasy feeling produced is not always acceptable. Repellent- or insecticide-impregnated (e.g. permethrin) clothing, such as wide-mesh jackets and hoods (as used by military personnel) give longer protection than repellents applied to the skin. If treated clothing is kept in plastic bags when not in use it should remain effective for many months before re-impregnation is needed.

Aerosols, mists and fogs

Hand-held (Swingfog, Dynafog) or vehicle-mounted (Leco, TIFA, Mirco-Gen) machines that generate insecticidal mists (51–100 μm), aerosols (<50 μm) or fogs (<15 μm) can be used to kill outdoor resting (exophilic) adult mosquitoes. Indoor resting (endophilic) adults are also occasionally killed by such fogging machines. Fogs are produced when very fine aerosol droplets are so numerous that they substantially reduce visibility. Several insecticides can be used, including malathion, bendiocarb (Ficam), fenitrothion (Sumithion), pirimiphos methyl (Actellic) and the synthetic pyrethroids. Although such applications can be spectacular there is very little residual effect and areas cleared of adult mosquitoes are rapidly invaded by newly emerging adults and mosquitoes flying in from outside the treated area. Repeated applications are needed to sustain control.

Applications of aerosols and mists are best made in calm weather, and usually in the evenings or early mornings when there are usually fewer thermals rising from the ground and less turbulence. Aerial applications from helicopters or fixed-wing aircraft usually give better coverage and more effective control than ground-based operations.

Ultra-low-volume applications Ultra-low-volume (ULV) techniques apply the minimum (<5 litres/ha) of concentrated insecticides, often just 225–500 ml/ha as against 5–25 litres/ha with conventional spraying. This means that trucks or aircraft can spray much larger areas on a tank of insecticides before they have to return to base. Insecticides commonly used in ULV operations include malathion, pirimiphos methyl (Actellic), fenitrothion (Sumithion), propoxur (Baygon), chlorpyrifos (Dursban) and the pyrethroids. With aerial applications, droplet size of the insecticide is bigger (150–200 μm) than that used in ground-based applications (50–100 μm) because they decrease in size, due to evaporation, as they fall to the ground. In addition to rapidly reducing outdoor resting and biting mosquito populations ULV spraying is used in potential or actual epidemic situations to control disease outbreaks. In emergency situations aerial spraying gives the fastest and most effective control of vectors, and has been used to stop transmission of haemorrhagic dengue, Japanese encephalitis and in North America the encephalomyelitis viruses.

Residual house-spraying

Some mosquitoes, such as many malaria vectors and *Culex quinquefasciatus*, rest in houses before and/or after blood-feeding. Their populations can be reduced by insecticidal spraying of houses, but as this approach is mainly used in malaria control operations it is described in Chapter 2.

FURTHER READING

Anon. (1995) Vector control without chemicals: Has it a future? A symposium. *Journal of the American Mosquito Control Association*, **11**, 247–93.

Bates, M. (1949) *The Natural History of Mosquitoes*. New York: Macmillan. (Facsimile by Harper Torchbooks, 1965.)

Bowen, M.F. (1991) The sensory physiology of host-seeking behavior in mosquitoes. *Annual Review of Entomology*, **36**, 139–58.

Chapman, H.C. (ed.) (1985) Biological control of mosquitoes. *American Mosquito Control Association Bulletin*, **6**, 1–218.

Clark, G.G. (coordinator) (1994) Prevention of tropical diseases: status of new and emerging vector control strategies. Proceedings of a symposium on vector control. *American Journal of Tropical Medicine and Hygiene*, **50** (Suppl.), 159pp.

Clements, A.N. (1992) *The Biology of Mosquitoes*, vol. 1, *Development, Nutrition and Reproduction*. London: Chapman & Hall.

Clements, A.N. (1999) *The Biology of Mosquitoes*, vol. 2, *Sensory Reception and Behaviour*. Wallingford: CABI Publishing.

Curtis, C.F. (ed.) (1989) *Appropriate Technology in Vector Control*. Boca Raton, Florida: CRC Press.

de Barjac, H. and Sutherland, D.J. (eds.) (1990) *Bacterial Control of Mosquitoes and Black Flies. Biochemistry, Genetics and Applications of* Bacillus thuringiensis *and* Bacillus sphaericus. New Brunswick: Rutgers University Press.

Gillett, J.D. (1971) *Mosquitos*. London: Weidenfeld & Nicolson.

Gillett, J.D. (1972) *Common African Mosquitos and their Medical Importance*. London: Heineman Medical.

Harbach, R. and Sandlant, G. (1997) *CABIKEY: Mosquito Genera of the World*. CD-ROM for Windows 3x or Windows 95 on a 486 PC with 8 MB RAM. Wallingford: CABI Publishing.

Horsfall, W.R. (1972) *Mosquitoes. Their Bionomics and Relation to Disease*. New York: Hafner Publishing.

Lacey, L.A. and Lacey, C.M. (1990) The medical importance of riceland mosquitoes and their control using alternatives to chemical insecticides. *Journal of the American Mosquito Control Association*, **6** (Suppl. 2), 1–93.

Laird, M. (1988) *The Natural History of Larval Mosquito Habitats*. London: Academic Press.

Laird, M. and Miles, J.W. (eds.) (1983) *Integrated Mosquito Control Methodologies*, vol. 1, *Experience and Components from Conventional Chemical Control*. London: Academic Press.

Laird, M. and Miles, J.W. (eds.) (1985) *Integrated Mosquito Control Methodologies*, vol. 2, *Biocontrol and Other Innovative Components, and Future Directions*. London: Academic Press.

Port, G.R., Boreham, P.F.L. and Bryan, J.H. (1980) The relationship of host size to feeding by mosquitoes of the *Anopheles gambiae* Giles complex (Diptera, Culicidae). *Bulletin of Entomological Research*, **70**, 133–44.

Ribeiro, J.M.C. (1987) Role of saliva in blood-feeding by arthropods. *Annual Review of Entomology*, **32**, 463–78.

Service, M.W. (1989) Rice, a challenge to health. *Parasitology Today*, **5**, 162–5.

Service, M.W. (1993) Mosquitoes (Culicidae). In *Medical Insects and Arachnids*, ed. R.P. Lane and R.W. Crosskey, pp. 120–40. London: Chapman & Hall.

Service, M.W. (1993) *Mosquito Ecology: Field Sampling Methods*, 2nd edn. London: Chapman & Hall.

World Health Organization (1973) Manual on larval control operations in malaria programmes. *WHO Offset Publication*, **1**, 1–199.

World Health Organization (1982) Manual on environmental management for mosquito control with special emphasis on malaria vectors. *WHO Offset Publication*, **66**, 1–238.

World Health Organization (1984) *Chemical Methods for the Control of Arthropod Vectors and Pests of Public Health Importance*. Geneva: World Health Organization.

World Health Organization (1990) *Equipment for Vector Control*. 3rd edn. Geneva: World Health Organization.

World Health Organization (1992) Vector resistance to pesticides. Fifteenth report of the WHO expert committee on vector biology and control. *World Health Organization Technical Report Series*, **818**, 1–71.

World Health Organization (1996) *Operational Manual on the Application of Insecticides for Control of the Mosquito Vectors of Malaria and Other Diseases*. WHO/CTD/VBC/96.1000. Geneva: World Health Organization.

World Health Organization (1997) *Vector Control. Methods for Use by Individuals and Communities*, prepared by J.A. Rozendaal. Geneva: World Health Organization.

See also references at the ends of Chapters 2 and 3.

2

Anopheline mosquitoes (Anophelinae)

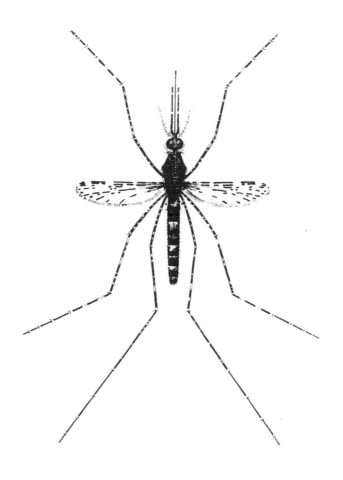

The subfamily Anophelinae contains three genera, but as explained in Chapter 1 only the genus *Anopheles* is of any medical importance. *Anopheles* mosquitoes have a world-wide distribution, occurring not only in tropical areas but also in temperate regions. There are about 430 different species. The most important disease carried by *Anopheles* mosquitoes is malaria. Some *Anopheles* species are also vectors of filariasis, especially that caused by *Wuchereria bancrofti*, and a few also transmit *Brugia malayi* and *Brugia timori*. A few species transmit arboviruses that are of minor medical importance.

2.1   EXTERNAL MORPHOLOGY OF *ANOPHELES*

The main features distinguishing adults of the Anophelinae, and in particular the genus *Anopheles*, from other mosquitoes have been given in Chapter 1, but a brief summary is presented here.

Most, but not all, *Anopheles* have spotted wings, that is the dark and pale scales are arranged in small blocks or areas on the veins (Fig. 1.12). The number, length and arrangement of these dark and pale areas differ considerably in different species and provide useful characters for species identification. Unlike Culicinae the dorsal and ventral surfaces of the abdomen are almost, or entirely, devoid of appressed scales. In both sexes the palps are about as long as the proboscis and in males, but not females, they are enlarged (that is clubbed) apically (Fig. 1.13).

2.2   LIFE CYCLE OF *ANOPHELES*

After mating and blood-feeding *Anopheles* lay some 50–200 small brown or blackish, boat-shaped eggs (Fig. 1.8) on the water surface. In most *Anopheles* there is a pair of conspicuous lateral, air-filled chambers called the floats on the eggs, which in a few species extend completely around the egg. These floats help to keep the eggs floating on the water surface. *Anopheles* eggs cannot withstand desiccation and in tropical countries they hatch within 2–3 days, but in colder temperate climates hatching may not occur until after about 2–3 weeks, the duration depending on temperature.

Anopheles larvae have a dark brown or blackish sclerotized head, a roundish thorax with numerous simple and branched hairs and a single pair of thoracic palmate hairs. The first six or seven abdominal segments usually have a pair of palmate hairs dorsally (Figs. 1.9, 1.10), which help maintain the larvae in a horizontal position at the water surface (Fig. 1.13). Segments 1–8 (segments 8 and 9 are combined) have a median sclerotized light or dark brown structure called a tergal plate, which varies in size and shape in different species. In addition each segment may possess one or two very small accessory tergal plates posterior to the main one (Fig. 1.9). Paired spiracles are present dorsally on the posterior end of the eighth

(8 + 9) visible segment; there is no siphon. On each side, just below and lateral to the spiracles, is a sclerotized structure bearing teeth, somewhat resembling a comb and called the pecten (Fig. 1.10). At the end of the last abdominal segment are four sausage-shaped transparent anal papillae which have an osmoregulatory function.

As in all mosquitoes there are four larval instars. *Anopheles* larvae are filter-feeders and unless disturbed remain at the water surface, feeding on bacteria, yeasts, protozoans and other micro-organisms, and also breathing in air through their spiracles. When feeding, larvae rotate their heads through 180° so that the ventrally positioned mouthbrushes can sweep the underside of the water surface. Larvae are easily disturbed by shadows or vibrations and respond by swimming quickly to the bottom of the water; they resurface some seconds or minutes afterwards.

Anopheles larvae occur in many different types of large and more or less permanent habitats, ranging from fresh- and salt-water marshes, mangrove swamps, grassy ditches, ricefields, edges of streams and rivers to ponds and borrow pits. They are also found in small and often temporary breeding places such as puddles, hoofprints, wells, discarded tins and sometimes water-storage pots. A few species occur in water-filled treeholes. In the Neotropical region (Central and South America and the West Indies) a few *Anopheles* breed in water that collects in the leaf axils of epiphytic plants growing on tree branches, such as bromeliads, which somewhat resemble pineapple plants. Some *Anopheles* prefer habitats with aquatic vegetation, others favour habitats without vegetation, some species like exposed sunlit waters whereas others prefer more shaded breeding places. In general *Anopheles* prefer clean and unpolluted waters and are usually absent from habitats that contain rotting plants or are contaminated with faeces.

In tropical countries the larval period frequently lasts only about 7 days, but in cooler climates the larval period may be about 2–4 weeks. In temperate areas some *Anopheles* overwinter as larvae and consequently may live many months.

In the comma-shaped pupa the head and thorax are combined to form the cephalothorax, which has a pair of short, trumpet-shaped breathing tubes, situated dorsally, with broad openings (Fig. 1.6). The abdominal segments have numerous short setae, and segments 2–7 or 3–7 have, in addition, distinct short, peg-like spines. The last segment terminates in a pair of oval paddles (Fig. 1.11).

Pupae normally remain floating at the water surface with the aid of the pair of palmate hairs on the cephalothorax, but when disturbed they swim vigorously down to the bottom with characteristic jerky movements. The pupal period lasts 2–3 days in tropical countries but sometimes as long as 1–2 weeks in cooler climates.

2.2.1 Adult biology and behaviour

Most *Anopheles* are crepuscular or nocturnal in their activities. Thus blood-feeding and oviposition normally occur in the evenings, at night or in the early mornings around sunrise. Some species such as *An. albimanus*, a malaria vector in Central and South America, bite people mainly outdoors (exophagic) from about sunset to 21:00 hours. In contrast, in Africa species of the *An. gambiae* complex, which contains probably the world's most efficient malaria vectors, bite mainly after 23:00 hours and mostly indoors (endophagic). As already discussed in Chapter 1, the times of biting, and whether adult mosquitoes are exophagic or endophagic, may be important in the epidemiology of diseases.

Both before and after blood-feeding some species will rest in houses (endophilic), whereas others will rest outdoors (exophilic) in a variety of natural shelters, such as amongst vegetation, in rodent burrows, in cracks and crevices in trees, under bridges, in termite mounds, in caves and among rock fissures, and cracks in the ground. Most *Anopheles* species are not exclusively exophagic or endophagic, exophilic or endophilic, but exhibit a mixture of these extremes of behaviour. Similarly, few *Anopheles* feed exclusively on either humans or non-humans, most feeding on both people and animals, but the degree of anthropophagism and zoophagism varies according to species. For example, *An. culicifacies*, an important Indian malaria vector, commonly feeds on cattle as well as humans, whereas in Africa *An. gambiae sensu stricto* (of the *gambiae* complex) feeds more rarely on cattle and thus maintains a stronger mosquito–man contact. This is one of the reasons why *An. gambiae* is a more efficient malaria vector than *An. culicifacies*.

2.3 MEDICAL IMPORTANCE

2.3.1 Biting nuisance

Although *Anopheles* mosquitoes may not be disease vectors in an area they may nevertheless constitute a biting nuisance. Usually, however, it is the culicine mosquitoes, especially *Aedes* and *Psorophora* species, that cause biting problems.

2.3.2 Malaria

Only mosquitoes of the genus *Anopheles* transmit the four parasites (*Plasmodium falciparum, P. vivax, P. malariae, P. ovale*) that cause human malaria. Because the sexual cycle of the malaria parasite occurs in the vector, it is conventional to call the mosquito the definitive host, and humans the intermediate host.

Male and female malaria gametocytes are ingested by female mosquitoes during blood-feeding and pass to the mosquito's stomach where they undergo cyclical development that includes a sexual cycle termed sporogony. Only gametocytes survive in the mosquito's stomach, all other blood forms of the malaria parasites (the asexual forms) being destroyed. Male gametocytes (microgametocytes) extrude flagella which are the male gametes (microgametes), and the process is called exflagellation. The microgametes break free and fertilize the female gametes (macrogametes) which have formed from the macrogametocytes. As a result of fertilization a zygote is formed, which elongates to become an ookinete. This penetrates the wall of the mosquito's stomach and reaches its outer membrane where it becomes spherical and develops into an oocyst. The nucleus of the oocyst divides repeatedly to produce numerous spindle-shaped sporozoites. When the oocyst is fully grown (about 40–80 μm) it ruptures and thousands of sporozoites are released into the haemocoel of the mosquito. The sporozoites (10–15 μm) are carried in the insect's haemolymph to all parts of the body but most penetrate the salivary glands.

The mosquito is now infective and sporozoites are inoculated into people the next time the mosquito bites. A single oocyst produces 1000 or more sporozoites, and it has been estimated that in heavy infections there may be as many as 60000–70000 sporozoites in the vector's salivary glands. However, the number may be much smaller than this, and very few (sometimes just 5–10) are actually injected into a person during feeding.

Oocysts can be seen on the stomach walls of vectors about 4–5 days after an infective blood-meal; after about 8 days they are fully grown and rupture. Sporozoites are usually found in the salivary glands after 8–13 days, but the time required for this cyclical development (extrinsic cycle), depends on both temperature and *Plasmodium* species. At 28 °C sporogony takes 8–16 days, at 20 °C 16 days and below 15 °C it cannot be completed.

The sporozoite rate, that is the percentage of female vectors with sporozoites in the salivary glands, varies considerably, not only from species to species of mosquito but also according to locality and season. Sporozoite rates are often about 1–5% in species such as *An. gambiae* and *An. arabiensis* of the *An. gambiae* complex, but less than 1% in many other species such as *An. albimanus* and *An. culicifacies*. For practical purposes it can be said that once a vector becomes infective it remains so throughout its life.

2.3.3 Important malaria vectors

Although there are over 400 species of *Anopheles* only about 70 are malaria vectors and of these probably only about 40 are important. Malaria vectors are often divided into primary and secondary vectors, but this can be misleading because a species may be classed as a primary vector in some areas but be only a secondary vector in others. Although presenting a list of important malaria vectors must be somewhat subjective I have nevertheless attempted to do this in Table 2.1 for the commonly recognized 12 epidemiological zones of malaria (Fig. 2.1).

Table 2.1. The main malaria vectors in the 12 epidemiological zones (see also Fig. 2.1)

1. North America (includes northern Mexico)	5. Mediterranean	10. Malaysia
An. freeborni	An. atroparvus	An. aconitus
An. quadrimaculatus	An. labranchiae	An. balabacensis
	An. sacharovi	An. campestris
2. Central America	An. superpictus	An. dirus
An. albimanus		An. donaldi
An. aquasalis	6. Afro-Arabian	An. flavirostris
An. darlingi	An. pharoensis	An. letifer
	An. sergentii	An. leucosphyrus
3. South America	7. Afrotropical	An. maculatus
An. albimanus	An. arabiensis	An. minimus
An. albitarsis	An. funestus	An. nigerrimus
An. aquasalis	An. gambiae	An. subpictus
An. darlingi		An. sundaicus
An. nuneztovari	8. Indo-Iranian	
An. pseudopunctipennis	An. culicifacies	11. Chinese
An. punctimacula	An. fluviatilis	An. anthropophagus
	9. Indo-Chinese hills	An. sinensis
4. North Eurasian	An. dirus	12. Australasian
An. atroparvus	An. fluviatilis	An. farauti
		An. koliensis
		An. punctulatus

Notes are given below on the principal larval habitats and biting behaviour of adults of the important malaria vectors listed in Table 2.1. These notes are no more than a guide to their behaviour, which may vary in different parts of the geographical distribution of a species. For convenience the vectors are listed alphabetically. Many species are found in species complexes (see p. 42) and some of the more important ones are referred to in the list of vectors; the most important and best known is the *An. gambiae* complex in Africa and more details are given of this complex than of the others.

Notes on malaria vectors

An. aconitus Ricefields, swamps, irrigation ditches, pools and streams with vegetation; prefers sunlit habitats. Adults feed indoors or outdoors on humans but also commonly on animals; adults rest indoors or outdoors after feeding.

An. albimanus Fresh or brackish waters such as pools, puddles, marshes, ponds and lagoons especially those containing floating or grassy vege-

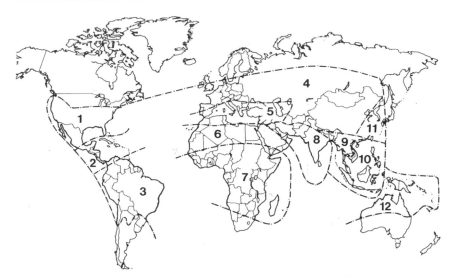

Figure 2.1 World map showing 12 malaria epidemiological zones (see also Table 2.1). 1, North American; 2, Central American; 3, South American; 4, North Eurasian; 5, Mediterranean; 6, Afro-Arabian; 7, Afrotropical; 8, Indo-Iranian; 9, Indo-Chinese hills; 10, Malaysian; 11, Chinese; 12, Australasian.

tation; prefers sunlit habitats. Adults feed on humans and domestic animals both indoors and outdoors; after feeding adults rest mainly indoors.

An. albitarsis One of four species in the *An. albitarsis* complex. Larvae nearly always in sunlit ponds, large pools and marshes with filamentous algae. Bites humans and domestic animals almost indiscriminately. Feeds outdoors and also indoors; usually rests outdoors after feeding.

An. anthropophagus Shaded pools and ponds, only rarely in ricefields. Bites humans indoors and mainly rests indoors after feeding.

An. aquasalis Tidal salt-water marshes, lagoons, salt-water regions of rivers, estuaries, rarely fresh water; sunlit or shaded habitats. Bites humans and domestic animals indoors or outdoors; rests mainly outdoors.

An. arabiensis One of seven species of the *An. gambiae* complex. See also under *An. gambiae*. Ricefields, borrow pits and also temporary waters such as pools, puddles, hoofprints; only in sunlit habitats. Adults bite humans indoors and outdoors but also cattle; after feeding rests either indoors or outdoors. Tends to occur in drier areas than does *An. gambiae*, and is more likely to bite cattle and rest outdoors than *An. gambiae*.

An. atroparvus One of 11 species in the *An. maculipennis* complex. Sunlit and exposed pools and ditches with either fresh or brackish water, rice-fields. Adults bite humans and domestic animals, and usually rest in stables, cow sheds and piggeries. Adults hibernate in these, and other shelters, during the winter, but periodically emerge to take blood-meals.

An. balabacensis Muddy and shaded forest pools, animal hoofprints and vehicle ruts, occasionally deep wells. Bites humans and cattle; feeds and rests outdoors. Morphologically and biologically very similar to *An. dirus*, but has a more restricted distribution – Sabah, Java, Borneo and certain Philippine islands.

An. bellator Although not listed in Table 2.1 *An. bellator* merits mentioning because its breeding places are atypical for anophelines. Larvae are found only in water collected in the leaf axils of bromeliads, which are epiphytes on trees in the Americas. Adults bite humans during the daytime in shaded forests, and also at night and may enter houses to feed. Adults rest mainly outdoors. The species will also bite domestic animals. It occurs in Trinidad, Venezuela, Surinam, Guyana and Brazil, where it can be a local malaria vector.

Another malaria vector that also breeds in bromeliad axils is *An. cruzii*, which is found in Central America and in Colombia, Peru, Venezuela and Brazil; its main importance as a vector is in coastal areas of Brazil.

An. campestris Deep and usually shaded or partially shaded waters such as ditches, wells, and shaded parts of ricefields; larvae also sometimes in brackish waters. Bites humans and animals indoors and outdoors; substantial numbers rest indoors after feeding.

An. culicifacies One of five species in the *An. culicifacies* complex. Great variety of clean and unpolluted habitats, irrigation ditches, pools, wells, borrow pits, edges of streams, marshes, ricefields and occasionally brackish waters. Prefers domestic animals but commonly bites humans indoors or outdoors; rests mainly indoors after feeding.

An. darlingi Fresh-water marshes, lagoons, ricefields, swamps, lakes, ponds, pools, edges of streams, especially with vegetation. Mainly shaded larval habitats. Feeds mainly on humans indoors; stays indoors after feeding.

An. dirus One of seven species in the *An. dirus* complex. Shaded pools, hoofprints in or at the edges of forests. Adults bite humans and domestic animals, mainly outdoors, and stay outdoors after feeding. Morphologically and biologically very similar to *An. balabacensis*, but has a more widespread distribution from western India to South-east Asia.

An. donaldi Shaded habitats such as tree-covered swamps, forest pools, often with vegetation, ricefields. Bites domestic livestock but also feeds on humans inside or outside houses; rests mainly outdoors.

An. farauti One of seven species in the *An. farauti* complex. Larvae usually occur in semipermanent waters such as swamps, ponds, lagoons, edges of slow-flowing streams, but also in puddles, hoofprints, pools and man-made containers; water may be fresh, slightly brackish or even polluted and in sunlight or shade. Adults bite animals but also humans indoors or outdoors; rest mainly outdoors, but also indoors.

An. flavirostris Flowing waters such as foothill streams, springs, irrigation ditches, also borrow pits and ricefields. Prefers shaded areas of sunlit habitats. Feeds mainly on humans, but also on domestic animals; feeds indoors or outdoors but rests mainly outdoors after feeding.

An. fluviatilis One of four species in the *An. fluviatilis* complex. Flowing waters such as hill streams, pools in river beds, irrigation ditches; prefers sunlight. Bites humans and domestic animals; feeds and rests either indoors or outdoors.

An. freeborni It has become apparent that this species consists of at least two separate but morphologically similar species, namely *An. freeborni* and *An. hermsi*. It seems that in the past malaria in California was transmitted by *An. hermsi*, not by *An. freeborni* as previously thought. It is possible that *An. hermsi* occurs in north-western Mexico and it, not *An. freeborni*, is the local malaria vector.

Larvae occur in ditches, irrigation ditches, pools, seepages, mostly with filamentous algae or floating or emergent vegetation, ricefields; prefers sunlight. Bites animals and also humans mainly outdoors but also indoors; rests indoors or outdoors. Overwinters as hibernating adults.

An. funestus More or less permanent waters, especially with vegetation, such as marshes, edges of streams, rivers and ditches, and ricefields with mature plants providing shade. Prefers shaded habitats. Bites humans predominantly, but also domestic animals; feeds indoors and also outdoors; after feeding rests mainly indoors.

An. gambiae One of seven species in the *An. gambiae* complex, comprising *An. gambiae*, *An. arabiensis*, *An. melas*, *An. merus*, *An. quadriannulatus*, *An. bwambae* and *An. quadriannulatus* species B, a newly described species from Ethiopia. These species can be separated by banding patterns of the polytene chromosomes found in the ovaries of half-gravid females and in the salivary glands of fourth-instar larvae. The species differ in certain aspects of their biology, behaviour, vector status and distribution.

Anopheles gambiae is the most anthropophagic species in the complex and the most important malaria vector – it is probably the world's most efficient vector. Larval habitats are virtually the same as for *An. arabiensis*, that is sunlit pools, puddles, borrow pits and ricefields. Bites humans both indoors and outdoors, and also feeds on domestic animals; rests mainly indoors, but will also rest outdoors.

Anopheles melas breeds in coastal salt waters such as mangrove swamps; it is restricted to West Africa. *Anopheles merus* is a coastal salt-water species in East and southern Africa, but can be found breeding in inland salt-water habitats. Adults of both species bite humans and rest indoors and outdoors; they are both regarded as secondary malaria vectors.

Anopheles quadriannulatus has a restricted distribution in eastern and southern Africa, it feeds almost exclusively on cattle and is not regarded as a malaria vector. *Anopheles bwambae* is restricted to breeding in warm mineral springs in Uganda; it is a rare species and not considered an important vector although locally it can transmit malaria.

See also under *An. arabiensis*.

An. koliensis Marshy pools, irrigation ditches, pools at edges of forest streams, often sunlit habitats. Adults bite humans and more occasionally animals, and after feeding rest mainly indoors, but they rest outdoors in some areas.

An. labranchiae One of 11 species in the *An. maculipennis* complex. Brackish waters in coastal marshes, fresh-water marshes, edges of grassy streams and ditches, ricefields; prefers sunlight. Bites humans and domestic animals indoors and outdoors; rests mainly in houses or animal shelters after feeding. Overwinters as hibernating adults.

An. letifer Often in acidic stagnant waters such as pools, swamps and ponds, especially on coastal plains; prefers shade. Bites livestock and humans mainly outdoors; rests outdoors.

An. leucosphyrus One of two species in the *An. leucosphyrus* complex. Clear seepage pools in forests. Adults bite humans inside and outside houses, but afterwards rest outdoors. This species is morphologically and biologically similar to *An. balabacensis* and *An. dirus*.

An. maculatus One of eight species in the *An. maculatus* complex. Seepage waters, pools formed in streams, edges of ponds, ditches and swamps with much vegetation; prefers sunlight. Bites humans and animals mainly outdoors and rests outdoors after feeding.

An. minimus One of four species in the *An. minimus* complex. Flowing waters such as foothill streams, springs, irrigation ditches, seepages, borrow pits and ricefields; prefers shaded areas. Feeds mainly on humans, but will bite domestic animals; feeds and rests mainly indoors.

An. nigerrimus Larvae in deep ponds, ricefields, irrigation ditches and marshes with much vegetation; prefers sunlight. Bites humans and animals mainly outdoors and rests mainly outdoors.

An. nuneztovari Possibly one of three species in an *An. nuneztovari* complex. Muddy waters of pools, vehicle tracks, hoofprints and small ponds, especially in and around towns; prefers sunlight. Feeds mainly on animals but in northern Colombia and western Venezuela bites humans indoors and outdoors, and rests outdoors after feeding.

An. pharoensis Marshes, ponds, especially those with abundant grassy or floating vegetation, also ricefields. Bites humans and animals indoors or outdoors; rests outdoors after feeding.

An. pseudopunctipennis Pools, puddles, seepage waters and edges of streams, especially habitats with algae; prefers sunlight. Feeds almost indiscriminately indoors or outdoors on humans and domestic animals; rests outdoors.

An. punctimacula Small pools, swamps, grassy pools at edges of streams; prefers shade. Bites humans and domestic animals both indoors and outdoors; rests indoors or outdoors after feeding.

An. punctulatus One of two species in the *An. punctulatus* complex. Temporary and often muddy pools, puddles, hoofprints and ditches. Bites humans in preference to animals; rests indoors or outdoors after feeding.

An. quadrimaculatus Ponds, marshes, borrow pits especially habitats with vegetation or filamentous algae, also in ricefields; prefers sunlight.

Bites humans and domestic animals indoors or outdoors, and rests indoors or outdoors after feeding. Overwinters as hibernating adults.

An. sacharovi Fresh or brackish waters of coastal or inland marshes, pools, ponds, especially those with vegetation. Prefers sunlit habitats. Bites humans and animals indoors or outdoors; usually rests in houses or animal shelters after feeding.

An. sergentii Borrow pits, ricefields, ditches, seepage waters, slow-flowing streams, sunlit or partially shaded habitats. Bites humans or animals indoors or outdoors; rests in houses and caves after feeding.

An. sinensis One of two species in the *An. sinensis* complex. Mainly rice-fields, but also marshes and grassy ponds. Adults bite humans and cattle; feeds indoors or outdoors; rests mainly outdoors but also found resting in houses. Overwinters as hibernating adults in cattle sheds, caves and other outdoor sites.

An. stephensi Possibly a species within a complex. Although not listed as a primary malaria vector in Table 2.1, *An. stephensi* can be an important vector locally, especially in towns. It is the main vector of urban malaria in much of the Indian subcontinent. Larvae breed in fresh, brackish or even polluted waters in man-made habitats such as water tanks, cisterns, wells, gutters, water-storage jars and containers. Adults bite humans indoors or outdoors, and rest mainly indoors afterwards. *Anopheles stephensi* has a very wide distribution, occurring from Iran and other countries in the Middle East across Pakistan, India, to Myanmar, Thailand and China.

An. subpictus One of four species in the *An. subpictus* complex. Muddy pools near houses, borrow pits, gutters, also brackish waters. Bites mainly animals, but also humans both indoors and outdoors; rests indoors or out-doors after feeding.

An. sundaicus One of three species in the *An. sundaicus* complex. Salt or brackish waters including lagoons, marshes, pools and seepages, especially with putrifying algae and aquatic weeds. Mainly a coastal species but found in fresh-water inland pools in Java and Sumatra. Prefers sunlit habitats. Bites humans and domestic animals indoors and outdoors; rests mainly indoors after feeding.

An. superpictus Flowing waters such as torrents of shallow water over rocky streams, pools in rivers, muddy hill streams, vegetation may be present; prefers sunlight. Bites humans and animals indoors and outdoors; after feeding rests mainly in houses and animal shelters, but also in caves.

2.3.4 Filariasis (details of vectors are given in Table 3.1)

Certain *Anopheles* species transmit filarial worms of *Wuchereria bancrofti*, *Brugia malayi* and *Brugia timori*, all of which cause filariasis in humans. The role of culicine mosquitoes as vectors of the first two species is described in Chapter 3.

Wuchereria bancrofti causes filariasis in people living in many tropical regions of the world (Central and South America, Africa and Asia, including the Pacific area), and also in some subtropical countries in the Middle East. In many of these areas bancroftian filariasis is mainly an urban disease. In contrast *B. malayi* is more a rural disease and has a much more restricted distribution, occurring in Asia in countries such as southern India, West Malaysia, Viet Nam, Indonesia, Thailand, China, New Guinea, Philippines and Korea. It is absent from Africa and the Americas.

Both bancroftian and brugian filariasis occur in two basic forms: nocturnal periodic and nocturnal subperiodic. Anophelines, together with certain culicines (*Culex quinquefasciatus*, various species of *Mansonia* and a few *Aedes* species), are the vectors of the nocturnal periodic form. Culicines are vectors of subperiodic forms (see Chapter 3).

In the nocturnal periodic form of these two parasites, most of the microfilariae during the day are in the blood vessels supplying the lungs. At night, especially during the middle part, microfilariae migrate to the peripheral blood system and lymph vessels. Because of this marked 24 hour periodicity microfilariae are ingested mainly by night-biting mosquitoes such as *Anopheles*.

The anophelines involved in filariasis transmission differ according to the area, but many of the principal malaria vectors are also important filarial vectors. For example, vectors of nocturnally periodic *W. bancrofti* include *An. aconitus, An. anthropophagus, An. aquasalis, An. balabacensis, An. dirus, An. flavirostris, An. funestus, An. gambiae, An. letifer, An. leucosphyrus, An. maculatus, An. nigerrimus, An. punctulatus, An. sinensis, An. subpictus* and *An. vagus*. There are no known animal reservoirs of the nocturnal periodic form of filariasis.

Nocturnal periodic *B. malayi* is transmitted by anophelines such as *An. anthropophagus, An. campestris, An. donaldi, An. barbirostris* and *An. sinensis*. There are no important animal reservoirs, although it is possible that some exist.

Brugia timori is known only from the small Indonesian islands of Timor, Flores, Alor and Roti. Its microfilariae are nocturnally periodic and are transmitted by *An. barbirostris*, and possibly other anopheline species. There are no known animal reservoirs.

Filarial development in mosquitoes

Filarial development of *W. bancrofti* and *B. malayi* within the mosquito vector, and the basic mode of transmission from mosquito to humans, are the same for all vectors. Basically, the life cycle in the mosquito is as follows. Microfilariae ingested with the blood-meal pass into the stomach of the mosquito (in some vectors such as *Anopheles* many may be destroyed during their passage through the oesophagus). Within a few minutes they exsheath, penetrate the stomach wall and pass into the haemocoel, from where they migrate to the thoracic muscles of the mosquito. In the thorax the small larvae become more or less inactive, grow shorter but considerably fatter and develop, after 2 days, into 'sausage-shaped' forms. They undergo two moults and the resultant third-stage larvae become active, leave the muscles and migrate through the head and down the fleshy labium of the proboscis. This is the infective stage and is formed some 10 days or more after the microfilariae have been ingested with a blood-meal.

When the mosquito takes further blood-meals, infective (third-stage) larvae (1.2–1.6 mm long) rupture the skin of the labella of the labium and crawl onto the surface of the host's skin. Several infective larvae may be liberated onto the skin when a vector is biting. However, many of these small worms die. Only a few manage to find a skin abrasion, sometimes the small lesion caused by the mosquito's bite, and thus enter the skin and pass to the lymphatic system. It should be noted that the salivary glands are not involved in the transmission of filariasis, and also that there is no multiplication or sexual cycle of the parasites in the mosquito.

Infection rates of infective larvae in anopheline vectors vary according to the mosquito species and local conditions, but they are often about 0.1–5% for *W. bancrofti* and about 0.1–3% for *B. malayi*.

There are no animal reservoirs of *W. bancrofti*, but the nocturnal subperiodic form of *B. malayi*, transmitted by *Mansonia* mosquitoes, is a zoonosis (see Chapter 3).

The presence of filarial worms in the thoracic muscles of mosquitoes, or infective worms in the proboscis, does not necessarily implicate mosquitoes as vectors of bancroftian or brugian filariasis. This is because there are several other mosquito-transmitted filariae. For example, various *Setaria* species infecting cattle, *Dirofilaria repens* and *Dirofilaria immitis* infecting dogs, and various other species of *Brugia*, such as *B. patei* in Africa and *B. pahangi* in Asia, infect animals but not people. Careful examination is therefore essential to identify the filarial parasites found in mosquitoes as those of *W. bancrofti* or *B. malayi*.

Table 3.1 (p. 72) summarizes the distribution and vectors of filariasis.

2.3.5   Arboviruses

The word arbovirus is derived from the term '*arthropod-borne virus*'. An arbovirus infection in either human or non-human hosts produces viraemia and the virus is ingested by blood-sucking insects such as mosquitoes when they take blood-meals. Within the vector the virus undergoes multiplication and/or cyclical development before being transmitted by the infected arthropod during refeeding. An arbovirus therefore undergoes obligatory development in an arthropod host; yellow fever and dengue are typical arboviruses transmitted by *Aedes* mosquitoes. In contrast the virus causing poliomyelitis is not an arbovirus for although it can be transmitted by certain flies, such as house-flies, this is purely mechanical transmission; the virus does not undergo any multiplication and/or development in an arthropod. The time taken for the infected mosquito to become infective, that is the extrinsic incubation period of the arbovirus, varies according to temperature, the species of arbovirus and mosquito. Over 510 arboviruses have been catalogued, but little is known about the epidemiology of many of them, and not all may be real arboviruses. About 100 arboviruses are transmitted by mosquitoes and infect humans. Many arboviruses are transmitted by culicine mosquitoes, in particular by *Aedes* and *Culex* species; other arboviruses are spread by ticks and other arthropods.

In 1959–60 a major epidemic of a painful but non-fatal disease called O'nyong nyong (a local African word meaning 'joint-breaker') was identified in Uganda and Kenya, and later in other countries of East and Central Africa. It was discovered that it was spread by the *An. gambiae* complex and *An. funestus*. This was the first time an *Anopheles* mosquito was incriminated with the spread of any arbovirus. About 20 other arboviruses infecting humans have since been found to be transmitted by *Anopheles*, some are also transmitted by culicine mosquitoes, but others are known only, or principally, from *Anopheles*.

2.4   CONTROL

The principal methods of mosquito control are described in Chapter 1. Here only those methods specifically directed against *Anopheles*, mainly because of their role as malaria vectors, are considered.

2.4.1   Larval control

Until the availability of DDT and other organochlorines in the mid-1940s, when spraying houses with insecticides became the main strategy in malaria control and malaria eradication programmes, larviciding with petroleum oils and Paris Green was widely carried out to reduce anopheline populations. When properly used, larvicides helped to reduce malaria transmission in localized areas of economic or social importance such as principal towns, coffee and tea estates, rubber plantations and mining camps. However, because of the repetitive nature (every 7–14 days) of this type of control it was logistically impossible to control malaria over large rural areas. This became possible only with the introduction of house-spraying. Nevertheless when larval habitats are readily identified larviciding, such as with temephos (Abate), may help reduce anopheline biting. The insect growth regulator pyriproxyfen in a trial in Sri Lanka showed promise in reducing malaria incidence, and the microbial insecticide *Bacillus thuringiensis* subsp. *israelensis* is favoured for malaria control in parts of India.

Certain predatory fish including *Gambusia* species continue to be used to reduce larval populations in some areas, but they have rarely resulted in significantly reducing disease transmission (p. 23).

Drainage of swamps and marshes has in some areas reduced the numbers of *Anopheles*, but drainage or filling-in of habitats is impractical in many situations, such as with species that breed in temporary habitats such as scattered pools and puddles, or those breeding in forest pools and swamps.

2.4.2   Adult control

In most malaria control campaigns control is now focused on the adults.

Residual house-spraying

The most widely practised method of residual house-spraying is the application of water-dispersible (wettable) powders of residual insecticides to the interior surfaces of walls, ceilings and roofs of houses. In the absence of resistance, DDT remains an effective insecticide; it is also relatively cheap and has a good safety record. However, because DDT accumulates in mammalian tissues and residues have been found in human breast milk many prefer not to use it, although the presence of residues does not necessarily mean that human health is affected. Reports that DDT causes breast cancer have not been substantiated. A well-recognised disadvantage of DDT is that it may be an irritant to mosquitoes, causing them to leave sprayed houses before they have picked up a lethal dose of insecticide. When used DDT is usually sprayed as a water-dispersible powder at the rate of 2 g/m^2. Even in highly endemic areas houses need be sprayed only at 6 monthly intervals, and where malaria transmission is very seasonal, such as just during the monsoon season, a single spraying before the rains may be sufficient to give good control.

If mosquitoes are resistant to DDT, or there are other reasons why DDT cannot be used, then organophosphates such as malathion, fenthion (Baytex), fenitrothion (Sumithion) or the carbamates, such as carbaryl (Sevin), bendiocarb (Ficam) or propoxur (Baygon) can be used, usually at rates of 1.5–2 g/m^2. Alternatively, lambdacyhalothrin (Icon) at 0.25–0.5 g/m^2 or permethrin at 0.5 g/m^2 can be used. However, these alternative insecticides are less persistent and spraying may have to be repeated at 3–4 monthly intervals. They are also more expensive than DDT. Furthermore, most are more toxic to people than DDT and so stricter safety measures have to be introduced during control programmes.

House-spraying often becomes popular because it kills bed-bugs, house-flies and cockroaches. However, the effectiveness of spraying houses in malaria control programmes depends on the mosquitoes resting indoors, and many important vectors, especially those in South-east Asia and the tropical Americas, feed and/or rest out of doors, and are thus not killed by residual house-spraying. There is also evidence that house-spraying may either promote the selection of exophilic populations of a species or favour the increase in numbers of a sibling species that is exophilic while reducing populations of a species of the complex that is primarily endophilic. Such population changes have been recorded in the *An. gambiae* complex in Zimbabwe, *An. nuneztovari* in Venezuela, *An. minimus* in Thailand and in a few other species.

Insecticide-impregnated bed-nets

Many of the bed-nets bought by poor communities are cheap, badly made and soon tear, and torn nets are useless in protecting people against mosquito bites. Also, mosquitoes will readily bite through a net if any part of the body is pressed up against it. However, nets that are impregnated with insecticides such as permethrin (200–500 mg/m^2), deltamethrin (25 mg/m^2) or lambdacyhalothrin (Icon) (10 mg/m^2), all of which repel and kill mosquitoes, will still protect people against bites even if they are torn or people sleep up against them. Nets can be impregnated by simply dipping them in insecticide contained in a plastic dustbin. Such impregnated nets should remain effective for 4–6 months before they need re-impregnating. If, however, lambdacyhalothrin is used for impregnation nets can remain effective for a year or more. Washing nets usually makes them less effective.

In Viet Nam and in China, where more than 2.5 million households are protected by impregnated nets, there have been significant reductions in malaria transmission and morbidity. In West Africa there have been reports of reductions in 'all-cause' child mortality in large-scale trials, but generally in Africa reductions in child mortality have been very variable and sometimes disappointing. It appears that impregnated nets usually protect only the users and do not have a mass-killing effect on vector populations in the area (e.g. village).

Nets will also give protection against filariasis, but will not be effective against vectors that bite early at night before people have gone to bed.

2.4.3   Malaria control and malaria eradication

In 1955 the eighth World Health Assembly stated that world-wide malaria eradication, except in Africa south of the Sahara, was technically feasible.

However, a sense of urgency in achieving this aim was recognized because insecticide resistance in *Anopheles* had been reported in 1950. In 1968 the 22nd World Health Assembly realized it had been overoptimistic and declared that global malaria eradication was not at present possible, although it remained the ultimate goal, and that for the time being malaria control should be the aim. This basically remains the situation today. The Global Malaria Control Strategy was initiated in 1992 at the Ministerial Conference on Malaria in Amsterdam and in 1995 the UN General Assembly endorsed a Special Initiative for Malaria Control in Africa. The latest development was the announcement in 1998 by the WHO of a global 'Roll Back Malaria' programme which will concentrate, at least initially, on Africa, and integrate with the African Initiative on Malaria.

The difference between malaria eradication and malaria control is that malaria eradication means the total cessation of transmission and elimination of the reservoir of infection in people so that at the end of the antimalaria campaign there is no resumption of transmission. Malaria control means reducing malaria transmission to an acceptable rate, that is to a level that no longer constitutes a major public health problem. This has the implication that control measures have to be maintained indefinitely; if they are relaxed malaria prevalence will rise. The feasibility of control will depend not only on scientific considerations but also on the financial and public health resources of the community, or country.

FURTHER READING

Beier, J.C. (1998) Malaria parasite development in mosquitoes. *Annual Review of Entomology*, **43**, 519–43.

Chavasse, D., Reed, C. and Attawell, K. (eds.), (1999) *Insecticide Treated Net Projects: A Handbook for Managers*. London and Liverpool: Malaria Consortium.

Collins, F.H. and Paskewitz, S.M. (1995) Malaria: current and future prospects for control. *Annual Review of Entomology*, **40**, 195–219.

Curtis, C.F. and Townson, H. (1998) Malaria: existing methods of vector control and molecular entomology. *British Medical Bulletin*, **54**, 311–25.

Gilles, H.M. and Warrell, D.A. (eds.) (1993) *Bruce-Chwatt's Essential Malariology*, 3rd edn. London: Edward Arnold.

Greenwood, B.M. (1997) What's new in malaria control? *Annals of Tropical Medicine and Parasitology*, **91**, 523–31.

Harrison, G. (1978) *Mosquitoes, Malaria and Man: A History of Hostilities Since 1880*. London: John Murray.

Jetten, T.H. and Takken, W. (1994) *Anophelism Without Malaria in Europe: A Review of the Ecology and Distribution of the Genus* Anopheles *in Europe*. Wageningen: Wageningen Agricultural University Papers, **94–5**.

Litsios, S. (1996) *The Tomorrow of Malaria*. Wellington, New Zealand: Pacific Press.

Molyneux, L. and Gramiccia, G. (1980) *The Garki Project: Research on the Epidemiology and Control of Malaria in the Sudan Savanna of West Africa*. Geneva: World Health Organization.

Nájera, J.A., Kouznetsov, R.L. and Delacollette, C. (1998) *Malaria Epidemics:*

Detection and Control Forecasting and Prevention. WHO/MAL/98.1084. Geneva: World Health Organization.

Oaks, S.C., Mitchell, V.S., Pearson, G.W. and Carpenter, C.J. (eds.) (1991) *Malaria: Obstacles and Opportunities*. Washington, DC: National Academy Press.

Rozendaal, J.A. (1989) Impregnated mosquito nets and curtains for self-protection and vector control. *Tropical Disease Bulletin*, **86**, R1–R41.

Sasa, M. (1976) *Human Filariasis: A Global Survey of Epidemiology and Control*. Tokyo: University of Tokyo Press.

Wernsdorfer, W.H. and McGregor, I. (eds.) (1988) *Malaria: Principles and Practice of Malariology*, vols. 1 and 2. Edinburgh: Churchill Livingstone.

World Health Organization (1975) Manual on practical entomology in malaria. Prepared by the WHO Division of Malaria and Other Parasitic Diseases. Part II. Methods and techniques. *WHO Offset Publication*, **13**, 1–191.

World Health Organization (1987) *Control of Lymphatic Filariasis: A Manual for Health Personnel*. Geneva: World Health Organization.

World Health Organization (1989) *Geographical Distribution of Arthropod-Borne Diseases and their Principal Vectors*. WHO/VBC/89.967: 3–134. Geneva: World Health Organization Vector Biology and Control Division.

World Health Organization (1992) Lymphatic filariasis; the disease and its control. *World Health Organization Technical Report Series*, **821**, 1–71.

World Health Organization (1992) *Entomological Field Techniques for Malaria Control. Part 1. Learner's Guide*. Geneva: World Health Organization.

World Health Organization (1992) *Entomological Field Techniques for Malaria Control. Part II. Tutor's Guide*. Geneva: World Health Organization.

World Health Organization (1993) *A Global Strategy for Malaria Control*. Geneva: World Health Organization.

World Health Organization (1993) Implementation of the global malaria control strategy. *World Health Organization Technical Report Series*, **839**, 1–57.

World Health Organization (1995) Vector control for malaria and other mosquito-borne diseases. *World Health Organization Technical Report Series*, **857**, 1–91.

Zahar, A.R. (1984–96) A series of ten World Health Organization mimeographed documents entitled 'Vector bionomics in the epidemiology and control of malaria', covering Europe, Africa, the southern and eastern Mediterranean regions, southwestern Arabia, Asia west of India, South-east Asia and the western Pacific region. This series is still in progress and will cover other regions.

See also references at the ends of Chapters 1 and 3.

3

Culicine mosquitoes (Culicinae)

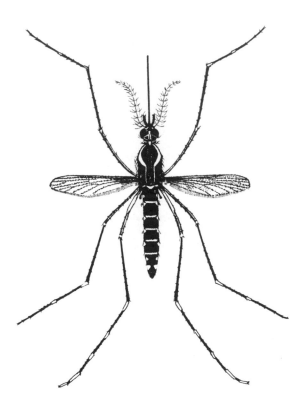

The subfamily Culicinae contains 33 genera of mosquitoes, of which the medically most important ones are *Culex*, *Aedes*, *Psorophora*, *Mansonia*, *Haemagogus* and *Sabethes*. The genera *Culex*, *Aedes* and *Mansonia* are found in both temperate and tropical regions, whereas *Psorophora* species are found only in North, Central and South America. *Haemagogus* and *Sabethes* mosquitoes are restricted to Central and South America.

Certain *Aedes* mosquitoes are vectors of yellow fever in Africa, and *Aedes*, *Haemagogus* and *Sabethes* are yellow fever vectors in Central and South America. *Aedes* species are also vectors of the classical and haemorrhagic forms of dengue. All six genera of culicine mosquitoes mentioned here, as well as some others, can transmit a variety of other arboviruses. Some *Culex*, *Aedes* and *Mansonia* species are important vectors of filariasis (*Wuchereria bancrofti* or *Brugia malayi*). *Psorophora* species are mainly pest mosquitoes.

Characters separating the subfamily Culicinae from the Anophelinae have been given in Chapter 1 and are summarized in Table 1.1.

It is not easy to give a reliable and non-technical guide to the identification of the six most important culicine genera. Nevertheless, characters that will usually separate these genera are given below, together with notes on their biology.

3.1   *CULEX* MOSQUITOES

3.1.1   Distribution
Culex mosquitoes are found more or less world-wide, but they are absent from the extreme northern parts of the temperate zones.

3.1.2   Eggs
The eggs are usually brown, long and cylindrical, laid upright on the water surface and placed together to form an egg raft which can comprise up to about 300 eggs (Fig. 1.15). No glue or cement-like substance binds the eggs to each other; adhesion is due to surface forces holding the eggs together. Eggs of a few other mosquitoes, including those of the genus *Coquillettidia*, also deposit their eggs in rafts.

3.1.3   Larvae
The larval siphon is often long and narrow (Fig. 3.1), but it may be short and fat. There is always more than one pair of subventral tufts of hairs on the siphon, none of which is near its base. These hair tufts may consist of very few short and simple hairs which may be missed unless larvae are carefully examined under a microscope.

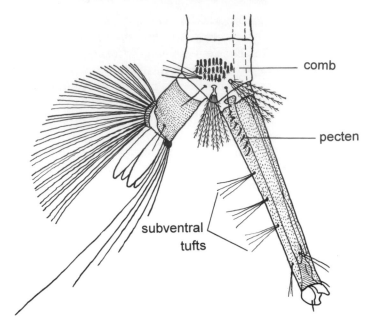

Figure 3.1 Terminal abdominal segments of a *Culex* larva showing the long siphon with three subventral tufts of hairs.

3.1.4   Adults

Frequently, but not always, the thorax, legs and wing veins of the adult are covered with sombre-coloured, often brown, scales. The abdomen is often covered with brown or blackish scales but some whitish scales may occur on most segments. Adults are recognized more by their lack of ornamentation than any striking diagnostic characters. The tip of the abdomen of females is blunt. The claws of all tarsi are simple and those of the hind tarsi are very small. Examination under a microscope shows that all tarsi have a pair of small fleshy pulvilli (Fig. 1.2).

3.1.5   Biology

Eggs are laid in a great variety of aquatic habitats. Most *Culex* species breed in ground collections of water such as pools, puddles, ditches, borrow pits and ricefields. Some lay eggs in man-made container-habitats such as tin cans, water receptacles, bottles and storage tanks. Only a few species breed in tree-holes and even fewer in leaf axils. The medically most important species, *Culex quinquefasciatus*, which is a filariasis vector, breeds in waters polluted with organic debris such as rotting vegetation, household refuse and excreta.

Larvae of this vector species are commonly found in partially blocked drains and ditches, soak-away pits, septic tanks and in village pots, especially the abandoned ones in which water is polluted and unfit for drinking. It is a mosquito that is associated with urbanization, and towns with poor and inadequate drainage and sanitation. Under these conditions its population increases rapidly.

Culex tritaeniorhynchus is an important vector of Japanese encephalitis and breeds prolifically in ricefields and also in grassy pools. In southern Asia larvae are not uncommon in fish ponds which have had manure added to them.

Culex quinquefasciatus, and many other *Culex* species, bite humans and other hosts at night. Some species, such as *C. quinquefasciatus*, commonly rest indoors both before and after feeding, but they also shelter in outdoor resting places.

3.2   *AEDES* MOSQUITOES

3.2.1   Distribution
World-wide, the range of *Aedes* mosquitoes extends well into northern and Arctic areas, where they can be vicious biters and serious pests to people and livestock.

3.2.2   Eggs
Eggs are usually black, more or less ovoid in shape and are always laid singly (Fig. 1.15). Careful examination shows that the eggshell has a distinctive mosaic pattern. Eggs are laid on damp substrates just beyond the water line, such as on damp mud and leaf litter of pools, on the damp walls of clay pots, rock-pools and tree-holes.

Aedes eggs can withstand desiccation, the intensity and duration of which varies, but in many species they can remain dry, but viable, for many months. When flooded, some eggs may hatch within a few minutes, while others of the same batch may require prolonged immersion in water; thus hatching may be spread over several days or weeks. Even when eggs are soaked for long periods some may fail to hatch because they require several soakings followed by short periods of desiccation before hatching can be induced. Even if environmental conditions are favourable, eggs may be in a state of diapause and will not hatch until this resting period is terminated. Various stimuli including reduction in the oxygen content of water, changes in daylength, and temperature may be required to break diapause in *Aedes* eggs.

Many *Aedes* species breed in small container-habitats (tree-holes, plant axils, etc.) which are susceptible to drying out; thus the ability of eggs to withstand desiccation is clearly advantageous. Desiccation and the ability of *Aedes* eggs to hatch in instalments can create problems with controlling the immature stages (p. 75).

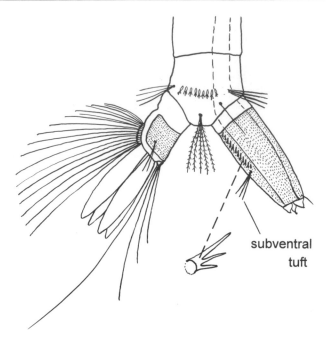

Figure 3.2 Terminal abdominal segments of an *Aedes* larva showing the short siphon with a single subventral hair tuft.

3.2.3 Larvae

Aedes species usually have a short barrel-shaped siphon, and there is only one pair of subventral tufts (Fig. 3.2) which never arises from less than one-quarter of the distance from the base of the siphon. Additional characters are at least three pairs of setae in the ventral brush, antennae that are not greatly flattened and no enormous setae on the thorax. These characters should separate *Aedes* larvae from most of the culicine genera, but not unfortunately from larvae of South American *Haemagogus*. In Central and South America *Aedes* larvae can usually be distinguished from those of *Haemagogus* by possessing either larger or more strongly spiculate antennae; also the comb is not on a sclerotized plate as in some *Haemagogus*.

3.2.4 Adults

Many, but not all, *Aedes* adults have conspicuous patterns on the thorax formed by black, white or silvery scales (Fig. 3.3); in some species yellow scales are present. The legs often have black and white rings (Figure 3.4*b*). *Aedes aegypti*, often called the yellow fever mosquito, is readily recognized by the lyre-shaped silver markings on the lateral edges of the scutum (Fig.

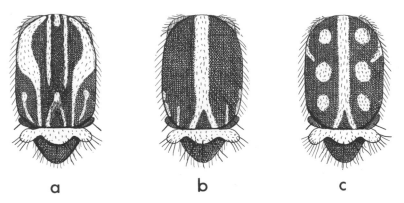

a   b   c

Figure 3.3 Dorsal surface of the thoraces of adult *Aedes* mosquitoes showing variations in pattern of black and white scales. (*a*) *Aedes aegypti* with typical lyre-shaped markings; (*b*) *Aedes albopictus*; (*c*) *Aedes vittatus*.

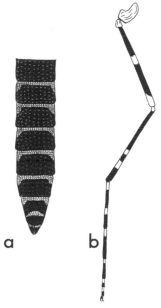

a   b

Figure 3.4 (*a*) Abdomen and (*b*) leg of an *Aedes* adult showing typical arrangements of black and white scales.

3.3*a*). Scales on the wing veins of *Aedes* mosquitoes are narrow, and are usually more or less all black, except that paler scales are often present at the base of the wings. In *Aedes* the abdomen is often covered with black and white scales forming distinctive patterns, and in the female it is pointed at its tip (Fig. 3.4*a*).

3.2.5   Biology

Although some *Aedes* species breed in marshes and ground pools, includ-ing snow-melt pools in arctic and subarctic areas, many, especially tropical species, are found in natural or man-made container-habitats such as tree-holes, bamboo stumps, leaf axils, rock-pools, village pots, tin cans and tyres. For example, *Ae. aegypti* breeds in village pots and water-storage jars placed either inside or outside houses. Larvae occur mainly in those with clean water intended for drinking. In some areas *Ae. aegypti* also breeds in rock-pools and tree-holes. *Aedes africanus*, an African species involved in the sylvatic transmission of yellow fever, breeds mainly in tree-holes and bamboo stumps, whereas *Ae. bromeliae*, another African yellow fever vector, breeds almost exclusively in leaf axils, especially those of banana plants, pineapples and coco-yams (*Colocasia*).

Aedes albopictus, which is a vector of dengue in South-east Asia, breeds in natural and man-made container-habitats such as tree-holes, water pots and vehicle tyres. This species was introduced into the USA in 1985 as dry, but viable, eggs which had been oviposited in tyres in Asia and then exported. By 1998 it had spread to 26 states in the USA, including Hawaii; and also been introduced by export of tyres to Brazil, possibly Bolivia, the Cayman islands, Colombia, possibly Cuba, the Dominican Republic, El Salvador, Guatemala, Mexico, Fiji, Papua New Guinea, Nigeria, Albania and Italy.

Larvae of *Ae. polynesiensis* occur in man-made and natural containers, especially split coconut shells, whereas larvae of *Ae. pseudoscutellaris* are found in tree-holes and bamboo stumps. Both these species are important vectors of diurnal subperiodic bancroftian filariasis. *Aedes togoi*, a vector of nocturnal periodic bancroftian and brugian filariasis, breeds principally in rock-pools containing fresh or brackish water.

The life cycle of *Aedes* mosquitoes from eggs to adults can be rapid, taking as little as about 7 days, but it more usually takes 10–12 days; in tem-perate species the life cycle may last several weeks to many months, and some species overwinter as eggs or larvae.

Adults of most *Aedes* species bite mainly during the day or early evening. Most biting occurs out of doors and adults usually rest out of doors before and after feeding.

3.3   *HAEMAGOGUS* MOSQUITOES

3.3.1   Distribution

Haemagogus mosquitoes are found only in Central and South America.

3.3.2   Eggs

The eggs are usually black and ovoid and laid singly in tree-holes and other natural container-habitats, occasionally in man-made ones. There is no

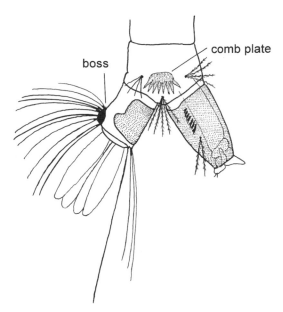

Figure 3.5 Terminal abdominal segments of a *Haemagogus* larva showing a ventral brush arising from a dark sclerotized boss, and comb scales arranged on a small 'plate'.

simple method of distinguishing eggs of *Haemagogus* from those of *Aedes* or *Psorophora* mosquitoes.

3.3.3 Larvae

Larvae have a single subventral tuft arising, as in *Aedes* larvae, not less than a quarter of the distance from the base of the siphon. They resemble *Aedes* larvae but can usually be distinguished by the antennae being short and either without spicules or with just a very few, and by a ventral brush arising from a sclerotized boss (Fig. 3.5). In some species the comb teeth are arranged at the edge of a sclerotized plate; in *Aedes* this plate is absent.

3.3.4 Adults

Adults are very colourful and can easily be recognized by the presence of broad, flat and bright metallic blue, red, green or golden coloured scales, covering the dorsal part of the thorax. Like *Sabethes* mosquitoes they have exceptionally large antepronotal thoracic lobes (Fig. 3.7a) behind the head. *Haemagogus* adults are rather similar to *Sabethes* in other respects, and it may be difficult for the novice to separate these two genera. However, no *Haemagogus* mosquito has paddles on the legs, which is a conspicuous feature of many, but not all, species of *Sabethes* (Fig. 3.7c).

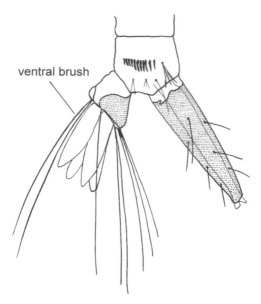

ventral brush

Figure 3.6 Terminal abdominal segments of a *Sabethes* larva showing a single pair of hairs in the ventral brush, numerous simple hairs on the siphon, and absence of a pecten on the siphon.

3.3.5   Biology

Eggs can withstand desiccation. Larvae occur mostly in tree-holes and bamboo stumps, but also in rock-pools, split coconut shells and sometimes in assorted domestic containers. They are basically forest mosquitoes. Adults bite during the day but mostly in the tree-tops where they feed on monkeys. Under certain environmental conditions, however, such as are experienced at edges of forests during tree felling operations or during the dry season, they may descend to the forest floor to bite humans and other hosts. *Haemagogus spegazzinii*, *Hg. equinus*, *Hg. leucocelaenus* (this species was previously placed in the genus *Aedes*), *Hg. janthinomys* and *Hg. capricornii* are all involved in yellow fever transmission in forest areas.

3.4   *SABETHES* MOSQUITOES

3.4.1   Distribution

Sabethes mosquitoes are found only in Central and South America.

3.4.2   Eggs

Little is known about the eggs of *Sabethes* species, but it appears that they are laid singly, have no prominent surface features such as bosses or sculpturing and are incapable of withstanding desiccation. The eggs of *Sabethes chloropterus*, a species sometimes involved in the sylvatic cycle of yellow

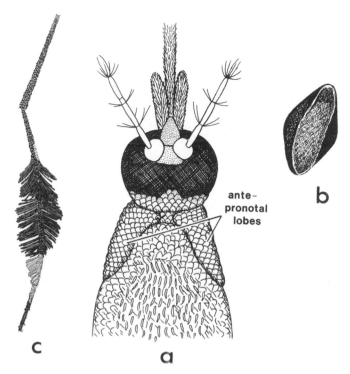

Figure 3.7 *Sabethes* mosquitoes. (*a*) Head and thorax showing antepronotal lobes forming a 'collar' behind the head; (*b*) an egg; (*c*) hind-leg showing narrow scales forming a paddle.

fever, are rhomboid in shape and can thus be readily identified from most other culicine eggs (Fig. 3.7*b*).

3.4.3   Larvae

The larval siphon has many hairs placed ventrally, laterally or dorsally, and is relatively slender and moderately long (Fig. 3.6). *Sabethes* larvae can usually be distinguished from other mosquito larvae by having only one pair of setae in the ventral brush, the comb teeth arranged in a single row, or at most with three or four detached teeth, and by the absence of a pecten.

3.4.4   Adults

The dorsal surface of the thorax of the adult is covered with appressed iridescent blue, green and red scales. The antepronotal lobes, like those in *Haemagogus*, are very large (Fig. 3.7*a*). Adults of many species have one or more pairs of tarsi with conspicuous paddles composed of narrow scales (Fig. 3.7*c*). Their presence immediately distinguishes *Sabethes* from all other mosquitoes. Species which lack these paddles resemble those of *Haemagogus* and a specialist is required to distinguish them.

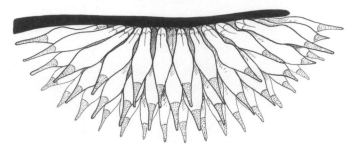

Figure 3.8 Eggs of *Mansonia* glued to the undersurface of floating vegetation.

3.4.5   Biology

Larvae occur in tree-holes and bamboo stumps; a few species are found in leaf axils of bromeliads and other plants. They are forest mosquitoes. They bite during the day, mainly in the tree canopy, but like *Haemagogus* adults may descend to ground level at certain times to bite humans and other hosts. *Sabethes chloropterus* has been incriminated as a sylvan vector of yellow fever.

3.5   *MANSONIA* MOSQUITOES

3.5.1   Distribution

Mansonia is principally a genus of wet tropical areas, but a few species occur in temperate regions.

3.5.2   Eggs

Eggs are dark brown–black and cylindrical, but have a tube-like extension apically which is usually darker than the rest of the egg. Eggs are laid in sticky compact masses, often arranged as a rosette, which are glued to the undersurface of floating vegetation (Figs. 1.15, 3.8).

3.5.3   Larvae

Mansonia larvae are very easily recognized because they have specialized siphons adapted for piercing aquatic plants (Fig. 3.9*b*, *c*) to obtain air. The siphon tends to be conical with the apical part darker and heavily sclerotized; the siphon has teeth and curved hairs which enable the larva to attach to plants and insert its siphon. Pupae also breathe through plants, by inserting their modified respiratory trumpets into them (Fig. 3.9*a*).

3.5.4   Adults

Typically adults have the legs (Fig. 3.10*c*), palps, wings and body covered with a mixture of dark (usually brown) and pale (usually white

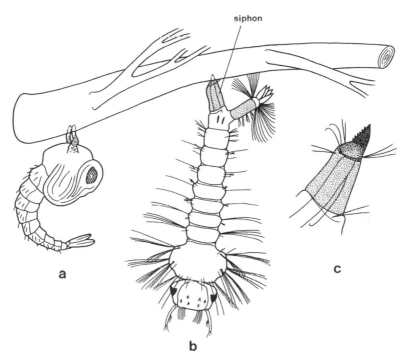

Figure 3.9 Immature stages of *Mansonia* mosquitoes. (*a*) Pupa with respiratory trumpets inserted into an aquatic plant; (*b*) larva with siphon inserted into a plant for respiration; (*c*) larval siphon showing serrated structures.

or creamy) scales, giving the mosquito a rather dusty appearance. The speckled pattern of dark and pale scales on the wing veins gives the wings the appearance of having been sprinkled with salt and pepper (Fig. 3.10*a*), and provides a useful character for identification. Closer examination shows that the scales on the wings are very broad and often asymmetrical (Fig. 3.10*b*). In other mosquitoes these scales are longer and narrower.

3.5.5 Biology

Eggs are glued to the undersurface of plants and hatch within a few days; they are unable to withstand desiccation. All larval habitats have aquatic vegetation, either rooted (such as grasses, rushes and reeds) or floating (such as *Pistia stratiotes*, *Salvinia* or *Eichhornia*). Larvae consequently occur in permanent collections of waters, such as swamps, marshes, ponds, borrow pits, grassy ditches, irrigation canals and even in the middle of rivers if they have floating plants.

Larvae and pupae only detach themselves from plants and rise to

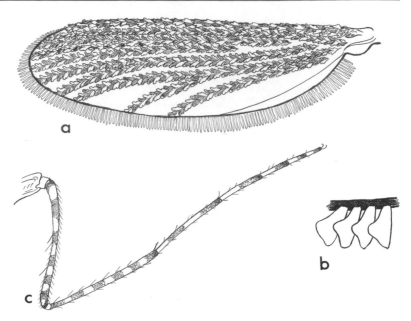

Figure 3.10 *Mansonia* mosquitoes. (*a*) Wing showing speckled distribution of dark and pale scales on the wing veins; (*b*) a few scales on a wing vein showing their broad shape; (*c*) leg showing distribution of dark and pale scales giving a banded pattern.

the surface of the water if they are disturbed. Because they are more or less permanently attached to plants the immature stages are frequently missed in larval surveys. It is therefore not easy to identify breeding places with certainty unless special collecting procedures are undertaken, such as the collection of plants to which the immature stages are thought to be attached. It is often difficult to control breeding of *Mansonia* species by conventional insecticidal applications, because of the problems of getting the insecticides to the larvae, which may be some distance below the water surface (see p. 77).

Adults usually bite during the night, but a few species are day-biters. After feeding, most *Mansonia* rest out of doors, but a few species rest indoors. The main medical importance of *Mansonia* mosquitoes is as vectors of filariasis, such as the nocturnal periodic and nocturnal subperiodic forms of *Brugia malayi* in Asia. In Africa filariasis (*W. bancrofti*) is not transmitted by *Mansonia* species, although they can be vectors of a few, but not very important, arboviruses.

3.6   *COQUILLETTIDIA* MOSQUITOES

The genus *Coquillettidia* is of minor medical importance and therefore described only briefly. It is mainly tropical but a few species occur in temperate regions. *Coquillettidia* is related to *Mansonia* and has sometimes been treated as a subgenus of *Mansonia*, but species of *Coquillettidia* differ in several respects. For example, as in *Culex* species, eggs are formed into egg rafts that float on the water, but they are narrower and longer than *Culex* rafts. Although larvae have rather conical siphons, which like those of *Mansonia* are inserted into plants for respiratory purposes, the larval antennae are much longer than those in *Mansonia*. Adults have narrow, not broad or heart-shaped, scales on the wings, and several species are a bright yellow. *Coquillettidia crassipes* can be a vector of nocturnal subperiodic *B. malayi* in Malaysia.

3.7   *PSOROPHORA* MOSQUITOES

Psorophora mosquitoes are also of only minor medical importance and so are described only briefly. They are found only in the Americas, from Canada to South America. They are similar in many respects to *Aedes* species. For example, their eggs look like those of *Aedes* and they can withstand desiccation, and a specialist is required to distinguish the larvae and adults of the two genera. Breeding places are mainly flooded pastures and sometimes ricefields; larvae of several species are predators. Although they can be vectors of a few arboviruses, such as Venezuelan equine encephalomyelitis, their main importance is as vicious biters; adults of some pest species can be very large.

3.8   MEDICAL IMPORTANCE

3.8.1   Biting nuisance

In several areas of the world a lot of money is spent on mosquito control, not so much because mosquitoes are vectors of disease but because they are such troublesome biters. For example, some of the best-organized mosquito control operations are in North America, where more money is spent on killing culicine mosquitoes than is expended in most tropical countries where they are important vectors of disease. In northern temperate and subarctic parts of America, Europe and Asia much greater numbers of *Aedes* mosquitoes can be encountered biting people than in tropical countries. Although they are not transmitting diseases to humans in these areas they can, nevertheless, make life outdoors almost intolerable.

3.8.2   Yellow fever

The arbovirus causing yellow fever occurs in Africa and tropical areas of the Americas. It does not occur in Asia or elsewhere, although mosquitoes capable of transmitting the disease occur in many countries. Yellow fever is a zoonosis, being essentially a disease of forest monkeys which under certain conditions can be transmitted to humans.

Africa

In Africa the yellow fever virus occurs in certain cercopithecid monkeys inhabiting the forests and is transmitted amongst them mainly by *Aedes africanus*. This is a forest-dwelling mosquito that breeds in tree-holes and bites mainly in the forest canopy soon after sunset – just in the right place at the right time to bite monkeys going to sleep in the tree-tops. This sylvatic, forest or monkey cycle, as it is sometimes called, maintains a reservoir in the monkey population (Fig. 3.11). In Africa, monkeys are little affected by yellow fever, dying only occasionally. Some species of monkeys involved in the forest cycle, such as the red-tailed guenon, descend from the trees to steal bananas from farms at the edge of the forest. In this habitat the monkeys get bitten by different mosquitoes including

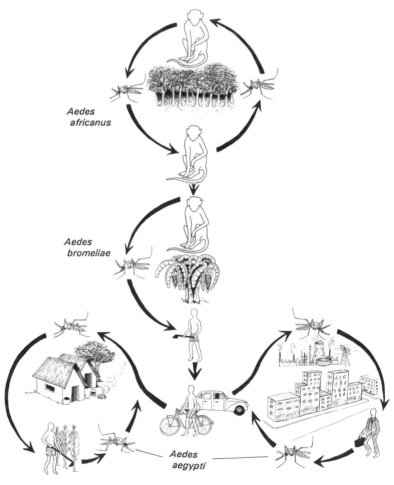

Aedes africanus

Aedes bromeliae

Aedes aegypti

Figure 3.11 Diagrammatic representation of the sylvatic, rural and urban transmission cycles of yellow fever in Africa.

Aedes bromeliae (formerly called *Ae. simpsoni*). This species bites during the day at the edges of forests and breeds in leaf axils of bananas, plantains and other plants such as coco-yams (*Colocasia*) and pineapples. If the monkeys have viraemia, that is yellow fever virus circulating in their peripheral blood, *Ae. bromeliae* becomes infected, and if the mosquito lives long enough it can transmit yellow fever to other monkeys or more importantly to people. This transmission cycle, occurring in clearings at the edge of the forest involving monkeys, *Ae. bromeliae*, and humans, is sometimes referred to as the rural cycle (Fig. 3.11). When people return to their villages they get bitten by different mosquitoes, including *Ae. aegypti*, a domestic species breeding mainly in man-made containers such as water-storage pots, abandoned tin cans and vehicle tyres. If people have viraemia then *Ae. aegypti* becomes infected and yellow fever is transmitted among the human population by this species. This is the urban cycle of yellow fever transmission (Fig. 3.11).

It is possible for people to become infected in the forest by bites of *Ae. africanus*, but the likelihood of humans acquiring yellow fever by a canopy-feeding mosquito is not very high. There is increasing evidence in West Africa that in rural areas other *Aedes* species spread the virus from monkeys to people. The epidemiology of yellow fever is complicated and variable. In some areas, for example, yellow fever may be circulating among the monkey population yet rarely gets transmitted to humans because local vector mosquitoes are predominantly zoophagic. Other primates in Africa such as bush-babies (*Galago* species) may also be reservoirs of yellow fever.

There is some evidence from West Africa that yellow fever virus may be trans-ovarially transmitted in *Aedes* species, as males have been found infected with the virus (see pp. 68–9 for an explanation).

Americas

In Central and South America the yellow fever cycle, although similar to that in Africa, differs in certain aspects (Fig. 3.12). As in Africa it is a disease of forest monkeys, mainly cebid ones (e.g. howler and spider monkeys), and is transmitted among them by forest-dwelling mosquitoes. The jungle vectors are various *Haemagogus* species including *Hg. equinus*, *Hg. spegazzinii*, *Hg. janthinomys*, *Hg. capricornii* and *Hg. leucocelaenus*. *Sabethes chloropterus* is also a vector although it seems to be a relatively inefficient one, and sometimes *Aedes* species are also involved in transmission. These are all arboreal mosquitoes which bite in the forest canopy and breed in tree-holes. New World monkeys are more susceptible to yellow fever than African monkeys and they frequently become sick and die. When people enter the jungle to cut down trees for timber, mosquitoes, which normally bite monkeys at canopy heights, may descend and bite them; if these

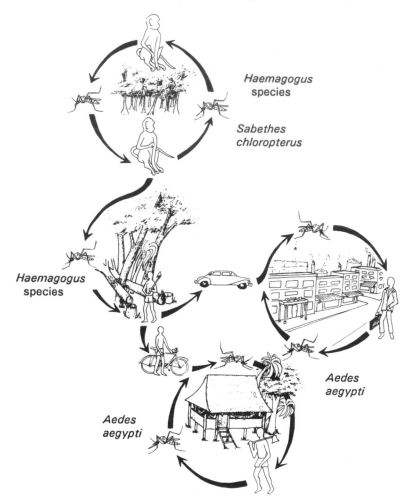

Figure 3.12 Diagrammatic representation of the jungle, rural and urban transmission cycles of yellow fever in Central and South America.

mosquitoes are infected people will develop yellow fever. The disease is then spread from person to person in villages and towns, as in Africa, by *Ae. aegypti* (Fig. 3.12). It is possible that in South America marsupials and even sloths and porcupines may be involved in yellow fever cycles.

In Panama it has been shown that infected female *Haemagogus* species can pass yellow fever virus through their ovaries to eggs and larvae and then to the next generation of adults, so that they are already infected although they have not as yet bitten a host. This phenomenon is called trans-ovarial transmission, and is more usually associated with tick-borne diseases. The importance of trans-ovarial transmission in the epidemiology of yellow fever is not clear, but it could perhaps maintain the virus during the dry season when there are few adult mosquitoes.

3.8.3 Viraemia

The intrinsic incubation period of yellow fever in humans is about 4–5 days, usually a little less in monkeys. Thus, after 4 or 5 days the virus appears in the peripheral blood, that is viraemia is produced, and this occurs irrespective of whether monkeys or humans are showing overt symptoms of the disease. Viraemia lasts only 2–3 days, after which the virus disappears from the peripheral blood never to return and the individual is immune. Monkeys and people are therefore infective to mosquitoes for only about 2–3 days in their entire lives. A relatively high titre of yellow fever (and also any other arbovirus) is needed before it can pass across the gut cells of the mosquito into the haemolymph, from where it invades many tissues and organs, including the salivary glands, where virus multiplication occurs. This is the extrinsic cycle of development and can take 5–30 days, depending on temperature, the type of virus and the mosquito species, but in yellow fever, and most other mosquito-borne viral infections, the incubation period is typically 12–15 days. A mosquito must therefore live a sufficiently long time before it becomes infective and capable of transmitting an arbovirus – or malaria or filariasis.

3.8.4 Dengue

A number of different viruses (dengue types 1, 2, 3 and 4) are responsible for dengue. Dengue is widely distributed in the tropics, occurring throughout most of South-east Asia, the Pacific, the Indian subcontinent, Africa, the USA down to northern parts of South America, and in the Caribbean. A more severe form, dengue haemorrhagic fever, causes infant mortality and has appeared in many parts of South-east Asia and also India. In 1981 haemorrhagic dengue and dengue type 4 were first noticed in Cuba. Dengue type 4 and haemorrhagic dengue have since been reported from at least 32 countries in the Americas, including many of the Caribbean islands. Dengue is now the most important viral disease transmitted by mosquitoes, having been recorded from more than 100 countries, and the numbers of cases world-wide are increasing.

Both dengue and haemorrhagic dengue are transmitted by *Ae. aegypti* and in South-east Asia to a lesser extent also by *Ae. albopictus*; both species breed in natural and man-made container-habitats such as water pots and tyres. Mosquitoes of the *Ae. scutellaris* group, which also breed in natural and man-made containers, may also be vectors in the Pacific islands. There are no known animal reservoirs, but in Malaysia there is transmission among monkeys by the *Ae. niveus* group of mosquitoes, larvae of which occur in tree-holes and bamboo stumps.

3.8.5   The encephalitis viruses

There are many other mosquito-borne arboviruses infecting humans in various parts of the world, for example Ross River virus (Australia), Chikungunya (Africa, Asia), Sindbis (Africa, Asia, Australia), Rift Valley fever (Africa) and West Nile (Africa, India, Europe). All these diseases are zoonoses. The most important arboviral diseases are the following encephalitis ones.

1. Japanese encephalitis (JE): occurs in Japan, China, Malaysia, Korea and other areas of South-east Asia and India. The basic transmission cycle involves birds, mainly herons, egrets and ibises. Pigs, however, are important reservoir hosts which develop high viraemias, and are called amplifying hosts. Transmission to birds, humans and pigs is mainly by *Culex tritaeniorhynchus*, which is a common ricefield breeding mosquito. *Culex gelidus*, breeding in streams and ricefields, probably maintains the virus in pig-to-pig transmissions in South-east Asia, whereas the *Cx. vishnui* group in India are vectors. Bats also become infected, and may play a role in the transmission cycle.

2. St Louis encephalitis (SLE): occurs in southern Canada and is widely distributed in the USA and extends down into Central and South America. In eastern USA it is mainly an urban disease spread by *Cx. quinquefasciatus* or *Cx. molestus*. Amplifying hosts are chickens and peridomestic wild birds. In rural areas *Cx. tarsalis* is a vector.

3. Eastern equine encephalomyelitis (EEE): occurs in Canada and mainly in the eastern area of the USA, but extends down to South America. It is probably the most severe encephalitis virus in humans and horses. EEE is principally an infection of birds, and in North America it is spread among them mainly by *Culiseta melanura* (the genus *Culiseta* is not discussed in this book). It is transmitted to people and horses, and sometimes also birds, by various *Aedes* species, such as *Ae. sollicitans* and *Ae. vexans*.

4. Western equine encephalomyelitis (WEE): widely distributed in the USA, but mainly in the western parts, it also extends into South America as far as Uruguay. This is basically an arboviral infection of birds which is transmitted by *Cx. tarsalis*, a ricefield breeding mosquito, as well as by other *Culex* species and *Culiseta melanura*. These vectors transmit the virus to birds, humans and horses.

5. Venezuelan equine encephalomyelitis (VEE): the infection is often fatal in horses but usually very mild in humans. VEE is found in southern USA through Central to northern parts of South America and is transmitted among birds, and occasionally bats, by several

species of *Culex*, *Aedes* and *Psorophora*. In Central and South America the virus also occurs in forest rodents, marsupials and monkeys. The basic cycle of transmission involves mammals rather than birds, and horses and people are commonly infected. Other mosquito genera, especially *Mansonia* species, have been incriminated in spreading VEE.

All these encephalitis viruses have a more complex epidemiology than described here, involving several different transmission cycles having different animal reservoirs and mosquito vectors. With the equine viruses horses are particularly susceptible and often die.

The viraemias produced in humans by JE, EEE, SLE, VEE and sometimes also by WEE, are so low that the infection cannot be transmitted by mosquitoes from humans to humans or from humans to other susceptible hosts. Thus humans are often referred to as 'dead-end' hosts. Similarly horses are 'dead-end' hosts for the encephalitis viruses infecting them.

3.8.6 Filariasis (Table 3.1)

The development in mosquitoes of filarial worms causing lymphatic filariasis is briefly described in Chapter 2 in connection with the role of anopheline vectors. Both bancroftian and brugian filariasis occur in two distinct forms. The nocturnal periodic form, in which the microfilariae are in the peripheral blood only at night, is transmitted by night-biting mosquitoes such as *Anopheles*, *Mansonia* and *Culex quinquefasciatus*. During the day the microfilariae are in the blood vessels supplying the lungs, and are not available to be taken up by mosquitoes. In subperiodic forms of *W. bancrofti* and *B. malayi* the microfilariae exhibit a reduced periodicity and are present in the peripheral blood during the day as well as at night, but there nevertheless remains a degree of periodicity. For example, subperiodic *W. bancrofti*, such as found in Polynesia, has a small peak in microfilarial density during the daytime and can therefore be called diurnal subperiodic, whereas subperiodic *B. malayi* in West Malaysia, Sumatra, Sabah, Thailand, etc., has a slight peak of microfilariae at night, and so can be called nocturnal subperiodic.

At least 38 *Anopheles* species and 35 culicine species have been incriminated as vectors of bancroftian and brugian filariasis, but only some of the more important species can be given in the following account and in Table 3.1.

Bancroftian filariasis

Wuchereria bancrofti occurs throughout much of the tropics (Central and South America, Africa, Asia, Pacific areas) and also in subtropical countries in the Middle East. It is the most widely distributed filarial infection of

Table 3.1. Summary of principal mosquito vectors of filariasis

Species and forms of filariasis	Geographic distribution	Vectors	Zoonotic reservoir
Wuchereria bancrofti			
Nocturnal periodic	Throughout tropics (but not Polynesia)	*Anopheles* spp., including many malaria vectors of different areas	None
	Throughout tropics (but not Polynesia)	*Culex quinquefasciatus*	
	New Guinea	*Mansonia uniformis*	
	China	*Aedes togoi*	
	Philippines	*Aedes poicilius*	
Diurnal subperiodic	Polynesia	*Aedes polynesiensis*	None
	Fiji	*Aedes pseudoscutellaris*	
	New Caledonia	*Aedes vigilax*	
Nocturnal subperiodic	Thailand	*Aedes niveus* group	None
Brugia malayi			
Nocturnal periodic (principally open swamps)	Asia, from India to Japan	*Anopheles* spp., including many malaria vectors of different areas	Not important, possibly some exist
	Asia, from India to Japan	*Mansonia annulifera, Ma., uniformis, Ma. annulata*	
	China, Korea and Japan	*Aedes togoi*	
Nocturnal subperiodic (mainly in swampy forests)	Malaysia, Indonesia, Thailand and Philippines	*Mansonia dives, Ma. bonneae, Ma. annulata, Ma uniformis, etc., Coquillettidia crassipes*	Monkeys, especially leaf-monkeys (*Presbytis* spp.), wild and domestic cats, and pangolins
Brugia timori			
Nocturnal period	Timor, Flores, Rotor and Alor islands in Indonesia	*Anopheles barbirostris*	None

humans. Bancroftian filariasis is essentially an urban disease, there are no animal reservoirs and the parasites develop in only mosquitoes and humans.

The **nocturnal periodic** form is transmitted by various *Anopheles* species (Chapter 2) and throughout much of its distribution also by *Culex quinquefasciatus*. This mosquito is widespread in the tropics and breeds mainly in waters polluted with human or animal faeces or rotting vegetation and other filth; larvae are found in septic tanks, cess pits, pit latrines, drains and ditches, and in water-storage jars if they contain organically polluted water. It is a mosquito that has increased in numbers in many towns due to rapid and increasing urbanization and the resultant proliferation of insanitary collections of water. Adults bite at night and after feeding they often rest in houses. Although *Cx. quinquefasciatus* is a good vector in most areas of Africa, it is a poor vector in West Africa where most transmission of bancroftian filariasis is by anopheline mosquitoes.

In New Guinea *Mansonia uniformis* and night-biting *Culex* species are also vectors.

In China, and formerly Japan, *Aedes togoi*, which breeds in rock-pools containing fresh or brackish water and in rain-filled receptacles such as pots and cisterns, is also a vector of nocturnal periodic bancroftian filariasis, although it is more usually known as a vector of brugian filariasis. Adults bite early in the evening around sunset.

In the Philippines *Ae. poicilius* is the most important vector. Adults bite in the early part of the night, mainly indoors but also sometimes outdoors. After feeding, adults rest outdoors. Larvae occur in leaf axils of banana, plantain and coco-yam (*Colocasia*) plants.

The **diurnal subperiodic** form occurs in the Polynesian region, from where the nocturnal periodic form is absent. The most important vector is *Ae. polynesiensis*, a day-biting mosquito which feeds mostly outdoors but may enter houses to feed; adults rest almost exclusively out of doors. Larvae occur in natural containers such as split coconut shells, leaf bracts and crab-holes, and also in man-made containers such as discarded tins, pots, vehicle tyres and canoes. *Aedes pseudoscutellaris* is another out of doors day-biting mosquito that is a vector of diurnal subperiodic *W. bancrofti* in Fiji. It breeds mainly in tree-holes and bamboo stumps but larvae are also found in crab-holes. In New Caledonia *Ae. polynesiensis* is absent and the most important vector is *Ae. vigilax*, adults of which feed outdoors mainly during the day. Larvae are found in brackish or fresh water in rock-pools and ground pools.

The **nocturnal subperiodic** form is found in Thailand and is transmitted by the *Ae. niveus* group of mosquitoes. Adults bite and rest outdoors; larvae are found mainly in bamboo.

It should be noted that although several *Aedes* mosquitoes are vectors of filariasis, especially the bancroftian form, *Ae. aegypti* is not a vector of lymphatic filariasis.

Natural infection rates of mosquitoes with infective larvae of *W. bancrofti* range from about 0.1% to 5%, depending greatly on vector species and local conditions.

Brugian filariasis

The **nocturnal periodic** form is principally a rural disease. The nocturnal form occurs throughout most of Asia, from southern India, West Malaysia, Viet Nam, Korea, Thailand, Indonesia, China to parts of Japan. It is transmitted mainly by night-biting *Mansonia* mosquitoes, such as *Ma. annulata*, *Ma. annulifera*, *Ma. uniformis* and also by various *Anopheles* species (Chapter 2). These *Mansonia* species breed in more or less permanent waters with floating or rooted aquatic vegetation, such as swamps and ponds. Adults bite mainly outdoors and rest out of doors after feeding, but they will also bite and rest indoors in some areas. In China, Korea and Japan *Ae. togoi* is a vector of nocturnal periodic *B. malayi*. There are no known important animal reservoirs.

The **nocturnal subperiodic** form of *B. malayi* occurs in West Malaysia, Indonesia, Thailand and the Philippines, and is transmitted by *Mansonia* mosquitoes, mainly by *Ma. annulata*, *Ma. dives*, *Ma. bonneae* and *Ma. uniformis*, and in Malaysia also by *Coquillettidia crassipes*. Larvae occur in habitats with much vegetation, such as swampy forests. Adults bite mainly at night but also during the daytime, especially species such as *Ma. dives* and *Ma. bonneae*. The subperiodic form of *B. malayi* is essentially a parasite of swamp monkeys, especially the so-called leaf monkeys (*Presbytis* species), humans becoming infected when they live at the edges of these areas. Other reservoirs include *Macaca* monkeys, domestic and wild cats, such as civets, and pangolins (Scaly anteaters, genus *Manis*).

Natural infection rates of mosquitoes with infective larvae of *B. malayi* range from about 0.1% to 2% or 3%, which is slightly lower than for *W. bancrofti*, but infection rates vary according to mosquitoes and local conditions.

As mentioned in Chapter 2, the discovery of filarial worms in mosquitoes does not necessarily imply they are vectors of either *W. bancrofti* or *B. malayi* because mosquitoes are also vectors of several other filarial parasites of animals.

3.9   CONTROL

The use of suitable repellents, mosquito nets and mosquito screening of houses and other personal protection measures (discussed in Chapters 1 and 2) can give some relief from culicine mosquitoes. It is, however, often more difficult to obtain protection from culicines than anophelines because many of them bite outdoors during the daytime. Spraying the interior surfaces of houses with residual insecticides, as practised for *Anopheles* control, is not usually applicable to culicines, because most species do not rest predominantly in houses. The main method of attack involves larval control, although sometimes aerial ultra-low-volume (ULV) applications are used to kill adult culicine mosquitoes.

3.9.1   *Aedes* and *Psorophora*

Control measures against mosquitoes such as *Ae. aegypti*, *Ae. polynesiensis* and *Ae. albopictus*, species which breed mainly in man-made containers in both rural and urban areas, are often aimed at reducing the numbers of larval habitats, that is control by source reduction. Thus, people are encouraged not to store water in pots inside or outside houses, or allow water to accumulate in discarded tin cans, bottles, vehicle tyres, etc. However, persuading people to co-operate in reducing peridomestic breeding of these vectors is often unsuccessful unless local legislation is strictly enforced. A reliable piped water supply to houses can do much to reduce *Ae. aegypti* breeding.

In situations where source reduction is not feasible, insecticidal applications to breeding sites can be attempted. Although insecticidal spraying will kill *Aedes* and *Psorophora* larvae it usually has no effect on their dry, but viable, eggs. These have been deposited at the edges of larval habitats and will hatch when the water level rises and floods them. *Aedes* and *Psorophora* mosquitoes breeding in pools, ponds and marshy areas can be controlled by ground-based or aerial applications of granular organophosphate insecticides. Applications can be made either before or after habitats have become flooded, that is pre- or post-flood treatments. Insecticidal granules landing on dry or muddy grounds remain more or less inactive until the habitats become flooded. When this occurs previously dry aedine eggs hatch, but at the same time flooding results in the release of insecticide from the granules and this kills the newly hatched larvae. This technique helps to overcome the problem of controlling *Aedes* and *Psorophora* mosquitoes whose eggs may hatch in instalments over extended periods after flooding.

Ground-based or aerial ULV applications of malathion, fenitrothion (Sumithion), pirimiphos-methyl (Actellic), propoxur (Baygon) or the synthetic pyrethroids may be the most appropriate control strategy in epidemic situations. Aeroplane ULV spraying has on several occasions helped to control *Ae. aegypti* during dengue epidemics, where one of the main objectives is to kill as rapidly as possible infected adult vectors so as to halt the epidemic.

When insecticides are used to kill mosquito larvae in water that is intended for drinking, it is essential that only those having extremely low mammalian toxicity are used; they should also impart no taste to the water. The insecticide that is usually recommended is temephos (Abate), in the form of 1% briquettes or sand granules that will give a concentration of 1 mg active ingredient/litre of water. However, some communities refuse to have their potable water dosed with any insecticide.

Yellow fever and dengue

Africa has over 90% of the world's number of yellow fever cases, and of the about 60 epidemics recorded in the past 70 years, 60% were in the last decade. The best defence against yellow fever is vaccination. In 1997 the WHO launched an appeal for a 5 year initiative to 'combat the dramatic resurgence of yellow fever in Africa', through mass immunization. Vector control, through sustained reduction in mosquito breeding, nevertheless still has a role in reducing the risks of yellow fever outbreaks; it also remains the main focus for dengue control, because at present there are no effective vaccines. Epidemics of dengue, and sometimes also yellow fever, may sometimes be curtailed by killing the adult vectors by ULV insecticidal spraying.

3.9.2   Culex

Culex quinquefasciatus, an important filariasis vector, is best controlled by improving sanitation and installing modern sewage systems, but often this is not feasible and insecticidal measures have to be employed. In most areas *Cx. quinquefasciatus* is resistant to a wide range of insecticides and this limits the choices of chemicals that can be used. Larval habitats should be sprayed every 7–10 days, and usually relatively large dosage rates are needed because most insecticides are less effective in the presence of organic pollution, which is characteristic of *Cx. quinquefasciatus* breeding places. Chlorpyrifos (Dursban) is usually one of the most effective insecticides in polluted waters. Synthetic pyrethroids and insect growth regulators such as methoprene (Altosid) and diflubenzuron (Dimilin) have also been used against *Cx. quinquefasciatus*.

Tipping non-toxic expanded polystyrene beads (2–3 mm) into pit latrines and cess pits to completely cover the water surface with a layer 2–3 cm thick prevents female *Cx. quinquefasciatus* from laying eggs in such breeding places. A single application can persist for several years and give excellent control. This is a control method that is readily accepted by most communities.

Insecticidal spraying of houses, as practised against malaria vectors, and the use of insecticide-impregnated bed-nets can be effective against several *Culex* species, including *Cx. quinquefasciatus*, because they are both endophilic and night-biters.

In the USA ULV spraying has frequently been used against various vectors of the encephalitis viruses.

Lymphatic filariasis

Control of lymphatic filariasis consists of treating the human host with microfilaricidal drugs (e.g. DEC, ivermectin) and concurrently reducing vector populations. Application of suitable larvicides including the microbial insecticides, *B. thuringiensis* subsp. *israelensis* and *B. sphaericus*, use of expanded polystyrene beads in pit latrines and septic tanks, use of insecticide-impregnated bed-nets, and possibly residual house-spraying can all reduce densities of *Culex quinquefasciatus*. In contrast it is very difficult to effectively reduce populations of *Mansonia* vectors, because of their often large and relatively inaccessible larval habitats, larval biology and because adults are not generally endophilic.

3.9.3 *Mansonia*

Mansonia mosquitoes are usually controlled by removing or killing the aquatic weeds upon which the larvae and pupae depend for their oxygen requirements. Weeds such as *Pistia stratiotes* can sometimes be removed manually from small areas such as borrow pits. Alternatively they can be sprayed with herbicides, such as diquat, 2,4-D, MCPA or pentachlorophenol (PCP). PCP has proved exceptionally good at killing *Salvinia*, an aquatic weed frequently associated with *Mansonia* breeding. Altering aquatic habitats by the removal of weeds may result in ecological changes that allow the habitats to become colonized by mosquito species that were previously excluded by the dense covering of weeds.

If insecticides are used to control *Mansonia* larvae, then granules or pellets are more suitable than liquid formulations. This is because they can penetrate the vegetation, sink to the bottom of breeding places and release their chemicals through the water. However, species such as *Ma. dives* and *Ma. bonneae*, which are important vectors of brugian filariasis, breed in extensive swampy forests, and are impossible to control because of the large and often inaccessible areas involved.

FURTHER READING

Anon. (1998) Main theme; dengue. *Public Health*, **14**, 6–58. Germany: Bayer A.G.

Chan, K.L. (1985) *Singapore's Dengue Haemorrhagic Fever Programme: A Case Study on the Successful Control of Aedes aegypti and Aedes albopictus Using Mainly Environmental Measures as a Part of Integrated Vector Control*. Tokyo: Southeast Asian Medical Information Center Publication no. **5**.

Gratz, N. and Knudsen, A.B. (1996) *The Rise and Spread of Dengue, Dengue Haemorrhagic Fever and its Vectors. A Historical Review (up to 1995)*. CTD/FIL(DEN) 96.7. Geneva: World Health Organization.

Gubler, D.J. and Kuno, G. (eds.) (1997) *Dengue and Dengue Haemorrhagic Fever*. Wallingford: CAB International.

Halstead, S.B. and Gomez-Dantes, H. (eds.) (1992) *Dengue: A World-wide Problem, A Common Strategy*. Proceedings of the international conference on dengue and *Aedes aegypti* community-based control. Mexico: Ministry of Health, Rockefeller Foundation.

Monath, T.P. (ed.) (1988) *Arboviruses: Epidemiology and Ecology*, vol. 1, *General Principles*, vol. 2, *African Horse Sickness to Dengue*, vol. 3, *Eastern Equine Encephalomyelitis to O'nyong nyong*, vol. 4, *Oropouche Fever to Venezuelan Equine Encephalomyelitis*, vol. 5, *Vesicular Stomatitis to Yellow Fever*. Boca Raton, Florida: CRC Press.

Nelson, M.J. (1986) Aedes aegypti: *Biology and Ecology*. PNSP/86–64. Pan American Health Organization mimeographed document.

Reeves, W.C. (ed.) (1990) *Epidemiology and Control of Mosquito-Borne Arboviruses in California, 1943–1987*. Sacramento, California: California Mosquito and Vector Control Association.

World Health Organization (1994) *Lymphatic Filariasis Infection & Disease: Control Strategies*. Report of a consultative meeting held at the Universiti Sains Malaysia, Penang, Malaysia (August 1994). TDR/CTD/FIL/Penang/94.1. Geneva: World Health Organization.

World Health Organization (1997) *Dengue Haemorrhagic Fever: Diagnosis, Treatment, Prevention and Control*, 2nd edn. Geneva: World Health Organization.

See also references at the ends of Chapters 1 and 2.

4

Blackflies (Simuliidae)

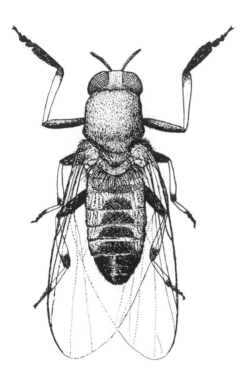

Blackflies have a world-wide distribution. There are nearly 1720 species in 26 genera. However, only four genera – *Simulium, Prosimulium, Austrosimulium* and *Cnephia* – contain species that bite people.

Medically, *Simulium* is by far the most important genus as it contains many vectors. In Africa, species in the *S. damnosum* complex and *S. neavei* group, and in Central and South America, species in the *S. ochraceum, S. metallicum* and *S. exiguum* complexes, transmit the parasitic nematode *Onchocerca volvulus* which causes human onchocerciasis (river blindness). In Brazil, *S. amazonicum* transmits *Mansonella ozzardi*, a filarial parasite that is usually regarded as non-pathogenic.

4.1   EXTERNAL MORPHOLOGY

The Simuliidae are commonly known as blackflies, but in some areas, in particular Australia, they may be called sandflies. As explained in Chapter 5 this latter terminology is confusing and best avoided because biting flies of the family Ceratopogonidae, and flies of the subfamily Phlebotominae, are also sometimes called sandflies.

Adult blackflies are quite small, about 1.5–4 mm long, relatively stout-bodied and, when viewed from the side, have a rather humped thorax. As their vernacular name indicates they are usually black in colour but many have contrasting patterns of white, silvery or yellowish hairs on their bodies and legs, and others may be predominantly or largely orange or bright yellow.

Blackflies have a pair of large compound eyes, which in females are separated on the top of the head (a condition known as dichoptic); in the males the eyes occupy almost all of the head, and touch on top of it and in front above the bases of the antennae (a condition known as holoptic). In the males, but not females, the small lenses are larger on the upper than lower half of the eyes (Fig. 4.1). The antennae are short, stout, cylindrical and distinctly segmented (usually 11 segments) but without long hairs. The mouthparts are short and relatively inconspicuous but the five-segmented maxillary palps, which arise at their base, hang downwards and are easily seen. Only females bite. The arrangement and morphology of the mouthparts are similar to those of the biting midges (Ceratopogonidae, Chapter 6). The mouthparts, being short and broad, do not penetrate very deeply into the host's tissues. Teeth on the labrum stretch the skin, while the rasp-like action of the maxillae and mandibles cuts through it and ruptures the fine blood capillaries. The small pool of blood produced is then sucked up by the flies. This method of feeding is ideally suited for picking up the microfilariae of *Onchocerca volvulus*, which occur in human skin not blood.

The thorax is covered dorsally with very fine and appressed hairs, which can be black, white, silvery, yellow or orange and may be arranged in various patterns. The relatively short legs are also covered with very fine

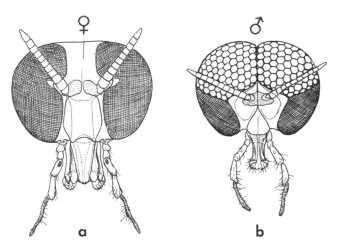

Figure 4.1 Front view of adult simuliid heads. (*a*) Female with dichoptic eyes; (*b*) male with holoptic eyes.

and closely appressed hairs and may be uni-colorous or have contrasting bands of pale and dark colour. The wings are characteristically short and broad and lack scales and prominent hairs. Only the veins near the anterior margin are well developed; the rest of the wing is membranous and has an indistinct venation (Fig. 4.2). The wings are colourless or almost so. When at rest the wings are closed over the body like the blades of a closed pair of scissors.

The abdomen is short and squat, and covered with inconspicuous closely appressed fine hairs. In neither sex are the genitalia very conspicuous. Blackflies are most easily sexed by looking at the eyes.

4.2   LIFE CYCLE

When first laid the eggs are pale and often whitish but soon darken to a brown or black colour. They are about 0.1–0.4 mm long, more or less triangular in shape but with rounded corners and have smooth unsculptured shells (Fig. 4.3*a*) which are covered with a sticky substance. They are always laid in flowing water but the type of breeding place differs greatly according to species. Habitats can vary from small trickles of water, slow-flowing streams, lake outlets and water flowing from dams to fast-flowing rivers and rapids. Some species prefer lowland streams and rivers whereas others are found in mountain rivers. In species such as *S. ochraceum*, one of the Central American vectors of onchocerciasis, eggs are scattered over the surface of flowing water while females are in flight. In most species, however, ovipositing females alight on partially immersed objects such as rocks, stones and vegetation to lay some 150–800 eggs in sticky masses or strings. Females may crawl underneath the water and become completely

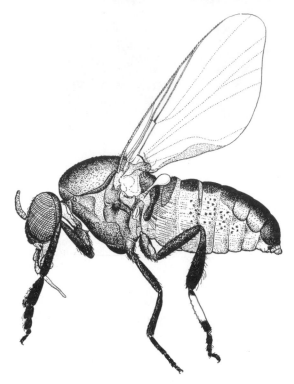

Figure 4.2 Adult simuliid (*Simulium damnosum*) in lateral view.

submerged during oviposition. There may be a few favoured oviposition sites in a stream or river, resulting in thousands of eggs from many females being found together. *Simulium damnosum*, for example, frequently has such communal oviposition sites.

Eggs of *S. damnosum* hatch within about 1–2 days but in many other tropical species the egg stage lasts 2–4 days. Eggs of species inhabiting temperate or cold northern areas may not hatch for many weeks and some species pass the winter as diapausing eggs.

There are six to nine (usually seven) larval instars and the mature larva is about 4–12 mm long, depending on the species, and is easily distinguished from all other aquatic larvae (Fig. 4.3*b*). The head is usually black, or almost so, and has a prominent pair of feeding brushes (cephalic fans), while the weakly segmented, cylindrical body is usually greyish, but may be darker or sometimes even greenish. The body is slightly swollen beyond the head and in most, but not all, species distinctly swollen towards the end. The rectum has finger-like rectal organs which on larval preservation may be extruded and visible as a protuberance from the dorsal surface towards the end of the abdomen. Ventrally, just below the head, is a small

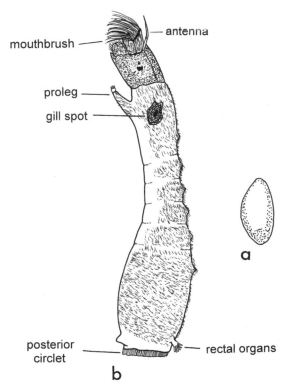

Figure 4.3 (a) Simuliid egg; (b) lateral view of the last larval instar of *Simulium damnosum* showing the body covered in minute dark setae and with dorsal tubercles.

pseudopod called the proleg which is armed with a small circlet of hooklets.

Larvae do not swim but remain sedentary for long periods on submerged vegetation, rocks, stones and other debris. Attachment is achieved by the posterior hook-circlet (anal sucker of many previous authors) tightly gripping a small silken pad. This has been produced by the larva's very large salivary glands and is firmly glued to the substrate. Larvae can nevertheless move about and change their position. This is achieved by alternately attaching themselves to the substrate by the proleg and the posterior hook-circlet; thus they move in a looping manner. When larvae are disturbed they can deposit sticky saliva on a submerged object, release their hold and be swept downstream for some distance at the end of a silken thread. They can then either swallow the thread of saliva and regain their original position, or reattach themselves at a site further downstream. Larvae normally orientate themselves to lie parallel to the flow of water with their head downstream. They are mainly filter-feeders, ingesting,

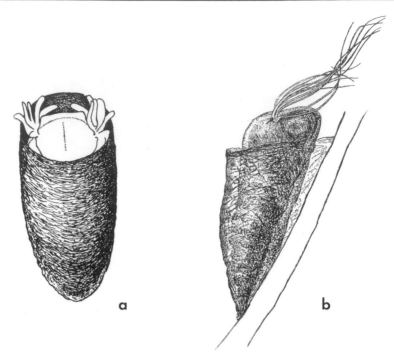

Figure 4.4 Simuliid pupae in cocoons. (*a*) Dorsal view of a species (*S. damnosum*) with broad and short respiratory filaments (by courtesy of The Natural History Museum, London); (*b*) lateral view of a species with long thin respiratory filaments.

with the aid of large mouthbrushes, suspended particles of food. However, a few species have predacious larvae and others are occasionally cannibal-istic. Larval development may be as short as 6–12 days depending on species and temperature, but in some species may be extended to several months, and in other species larvae overwinter.

Mature larvae, which can be recognized by a blackish mark termed the gill spot (respiratory organ of the future pupa) on each side of the thorax (Fig. 4.3*b*), spin, with the silk produced by the salivary glands, a protective slipper-shaped brownish cocoon. This cocoon is firmly stuck to submerged vegetation, rocks or other objects (Fig. 4.4) and its shape and structure vary greatly according to species. After weaving the cocoon the enclosed larva pupates. The pupa has a pair of, usually prominent, filamentous or broad thin-walled, respiratory gills. Their length, shape and the number of fila-ments or branches provide useful taxonomic characters for species identifi-cation. These gills, and the anterior part of the pupa, often project from the entrance of the cocoon (Fig. 4.4). In both tropical and non-tropical countries the pupal period lasts only 2–6 days and is unusual in not appearing to be dependent on temperature. On emergence adults either rise rapidly to the water surface in a protective bubble of gas, which prevents them from being wetted, or they escape by crawling up partially submerged objects such as vegetation or rocks. A characteristic of many species is the more or less simultaneous mass emergence of thousands of adults. On reaching the water surface the adults immediately take flight.

The empty pupal cases, with gill filaments still attached, remain enclosed in their cocoons after the adults have emerged and retain their taxonomic value. Consequently, they provide useful information on the species of simuliids that have recently bred and successfully emerged from various habitats.

A few African and Asian blackfly species have a very unusual aquatic existence. For example, in East Africa larvae of *S. neavei* (except first instar) and pupae do not occur on submerged rocks or vegetation but on other aquatic arthropods, such as the bodies of immature stages (nymphs) of mayflies (Ephemeroptera), and various crustaceans including fresh-water crabs. Such an association is termed a phoretic relationship. Eggs, however, are never found on these animals; they are probably laid on submerged stones or vegetation (p. 88).

The nuclei of the larval salivary gland cells have large polytene chromosomes which have banding patterns that are used to identify otherwise morphologically identical species within a species complex. For example, chromosomal studies have shown that there are about 40 cytologically different entities in the *S. damnosum* complex, several of which are distinct species, whereas other forms are best referred to as cytoforms until their reproductive isolation, or otherwise, is determined.

4.2.1   Adult behaviour

Both male and female blackflies feed on plant juices and naturally occurring sugary substances, but only females take blood-meals. Biting occurs out of doors at almost any daylight hour, but many species, including *S. damnosum*, seem to have bimodal biting patterns, with a peak in the early morning and another in the afternoon or early evening. However, in some species, such as *S. ochraceum* in Guatemala, biting continues more or less throughout much of the day. Many species seem particularly active on cloudy, overcast days and in thundery weather. Species may exhibit marked preferences for feeding on different parts of the body; for example, *S. damnosum* feeds mainly on the legs whereas *S. ochraceum* prefers to bite the head and torso. When feeding on animals, adults crawl down the fur of mammals, or feathers of birds, to bite the host's skin; they may also enter the ears to feed.

Many species of blackfly feed almost exclusively on birds (ornithophagic) and others on non-human mammalian hosts (zoophagic).

However, several species also bite people. Some human-biting species seem to prefer various large animals such as donkeys or cattle and bite humans only as a poor second choice, whereas others appear to find humans almost equally attractive hosts; no species bites people alone. In many species sight seems important in host location but host odours may also be important. After feeding, blood-engorged females shelter and rest in vegetation, on trees and in other natural outdoor resting places until the blood-meal is completely digested. In the tropics this takes 2–3 days, while in non-tropical areas it may take 3–8 days or longer, the speed of digestion depending mainly on temperature. A few species are autogenous. Relatively little is known about blackfly longevity, but it seems that adults of most species live for 3–4 weeks.

Females of some species may fly considerable distances (15–30 km) from their emergence sites to obtain blood-meals and may also be dispersed long distances by winds. For example, it is not exceptional for adults of *S. damnosum* to be found biting 60–100 km from their breeding places, and in West Africa there is evidence that prevailing winds can carry adults up to 400–600 km. The long distances involved in dispersal have great relevance in control programmes, because areas freed from blackflies can be reinvaded from distant breeding places.

In temperate and northern areas of the Palaearctic and Nearctic regions biting nuisance from simuliids is seasonal, because adults die in the autumn and new generations do not appear until the following spring or early summer. Although in many tropical areas there is continuous breeding throughout the year, there may nevertheless be dramatic increases in population size during the rainy season.

4.3   MEDICAL IMPORTANCE

4.3.1   Annoyance
In both tropical and non-tropical areas of the world blackflies can cause a very serious biting problem, since their bites can be painful. Although the severity of the reaction to bites differs in different individuals, localized swelling and inflammation frequently occurs, accompanied by intense irritation lasting for several days or even weeks. Repeated biting by blackflies such as *S. erythrocephalum* in central Europe, *S. posticatum* in England, *S. venustum* and *S. vittatum* in North America can cause headaches, fevers, swollen lymph glands and aching joints. In some areas of Canada outdoor activities are almost impossible at certain times of the year due to the intolerable numbers of biting simuliids. The classical example of the nuisance that can result from blackflies was the seasonal exodus during the eighteenth century of people from the Danube valley area in central Europe, largely to save their livestock from attack by the enormous numbers of *S.*

colombaschense. However, it is as disease vectors that blackflies are most important medically.

4.3.2   Onchocerciasis

Onchocerciasis is a non-fatal disease, often called river blindness, that is caused by the filarial parasite *Onchocerca volvulus*. There are no animal reservoirs, so the disease is not a zoonosis. Onchocerciasis occurs throughout West Africa, Central Africa, and much of East Africa from Ethiopia to Tanzania with isolated pockets of infection in Malawi, Sudan and southern Yemen, possibly extending into Saudi Arabia. About 95% of all cases occur in Africa. In the Americas onchocerciasis is localized in areas of southern Mexico, Guatemala, Brazil, Venezuela, Ecuador and Colombia.

Blackflies are the only vectors of human onchocerciasis. Their habit of tearing and rasping the skin to rupture the blood capillaries when obtaining a blood-meal makes them particularly suited to the ingestion of the skin-borne microfilariae of *O. volvulus*. Many of the microfilariae ingested during feeding are destroyed or excreted, but some penetrate the stomach wall and migrate to the thoracic muscles where they develop into sausage-shaped stages and undergo two moults. A few survive and elongate into thinner worms which pass through the head and down the short proboscis. The infective third-stage worms (about 660 μm in length) in the proboscis penetrate the host's skin when females alight to feed. The interval between the ingestion of microfilariae to the time infective larvae are in the proboscis is about 6–13 days, depending on temperature.

African vectors of onchocerciasis

Chromosomal studies have shown that the *S. damnosum* complex is composed of about 40 cytoforms, several of which merit species status. The *damnosum* complex is widespread in tropical Africa and certain of its species are the most important vectors of onchocerciasis in Africa.

Adults of the *damnosum* complex are mainly black. They can be recognized by the front tarsi being broad and flattened, having a conspicuous dorsal crest of fine hairs, and by the presence of a very broad white area on the first segment of the hind tarsus (basitarsus) (Fig. 4.2). Larvae are more or less covered with fine setae and there are usually prominent dorsal abdominal tubercles (Fig. 4.3*b*). The branches of the pupal respiratory gills are thick and finger-like and contained within the neck of the cocoon (Fig. 4.4*a*). The immature stages are found in the rapids of small or very large rivers in both savannah and forested areas of Africa. Adults frequently disperse hundreds of kilometres from their breeding sites. The vectorial status of species within the *damnosum* complex varies, the most important vectors of onchocerciasis being *S. damnosum s. str., S. sirbanum, S. sanctipauli* and *S. leonense.* Other, but much less important, vectors include *S. neavei* which is responsible for transmission in the Congo, Zaire and Uganda, and formerly in Kenya, where it was eradicated in the 1950s by bush clearing and insecticidal dosing of streams. *Simulium neavei* is a phoretic species, that is the larvae and pupae are found attached to other freshwater fauna, in this instance crabs of the genus *Potamonautes* and mayfly (Ephemeroptera) nymphs, both of which occur in small rocky, rather turbid streams and rivers.

American vectors of onchocerciasis

Simulium ochraceum is the principal onchocerciasis vector in southern Mexico and Guatemala, and is widely distributed in Central America and northern parts of South America. Adults are very small and are easily recognized by their dark brown legs, bright orange scutum (dorsal surface of the thorax) and yellow basal part of the abdomen, which contrasts with the black apical part. Females oviposit while in flight, dropping their eggs onto floating vegetation. Larval habitats consist of trickles of flowing water and very small streams which are often concealed by bushes, vegetation and fallen leaves. Adults do not appear to disperse far. The main biting season is unusual in being in the drier months of the year.

Simulium metallicum occurs in Mexico through Central America to northern areas of South America. In Venezuela it appears to be the most important vector whereas in other areas such as in Mexico and Guatemala it is considered a minor vector. It is a black species and has a broad white area on the first segment of the hind tarsus. Larvae occur in small or large streams and rivers. Adults fly further from their breeding sites than do those of *S. ochraceum.*

Simulium exiguum is the only known vector in Colombia, is a primary vector in Ecuador and a secondary one in Venezuela. In Brazil one of the main vectors is *S. guianense.*

4.3.3   *Mansonella ozzardi*

Mansonella ozzardi is a filarial parasite of humans that is usually regarded as non-pathogenic, although it has been reported as causing morbidity in Colombia and Brazil. It is transmitted in Central and South America by *Culicoides* species (Chapter 5), but in Brazil *S. amazonicum* is a vector.

4.4   CONTROL

There are few reports on the effectiveness of repellents against blackfly attacks. It seems that some protection, usually lasting up to 2 hours, can be gained by use of repellents such as diethyltoluamide (deet), dimethylphthalate (dimp) and butyryl-tetrahydro quinoline.

Although insecticidal fogging or spraying of vegetation thought to harbour resting adult blackflies has occasionally been undertaken, this approach results in very temporary and localized control. The only practical method at present available for the control of blackflies is the application of insecticides to their breeding places to kill the larvae. Insecticides need be applied to only a few selected sites on watercourses for some 15–30 minutes, because as the insecticide is carried downstream it kills simuliid larvae over long stretches of water. The flow rates of the water and its depth are used to calculate the quantity of insecticide to be released. Applying DDT by this method has in the past resulted in good control of *S. damnosum* in areas of Nigeria and Uganda. If treatment is not repeated at intervals throughout the year, gravid female adults dispersing into the area from untreated areas will probably cause recolonization. In Kenya *S. neavei* has been eradicated by the application of DDT to the relatively small streams in which this vector occurred, together with bush clearing. Because of its accumulation in food chains DDT is no longer recommended for larval control; instead organophosphates, such as temephos (Abate), are usually recommended.

In many areas ground application of larvicides is difficult, either because of the enormous size of the rivers requiring treatment or because breeding occurs in a large network of small streams and watercourses. Under these conditions aerial applications from small aircraft or helicopters have been used. Considerable information has been gained in North America on the chemical control of pestiferous blackflies.

4.4.1 Onchocerciasis control programme (OCP)

Because of the severity of river blindness in the Volta River Basin area of West Africa and its devastating effect on rural life the world's most ambitious and largest vector control programme was initiated in 1974 by the World Health Organization, and is called the Onchocerciasis Control Programme (OCP). This programme originally involved seven West African countries (Benin, Burkina Faso, Côte d'Ivoire, Ghana, Mali, Niger and Togo), but in 1986 with the addition of parts of Guinea, Guinea Bissau, Senegal and Sierra Leone the programme covered 11 countries. Some 50000 km of rivers over an area of 1.3 million km^2 that were breeding the *S. damnosum* complex were dosed at weekly intervals with temephos (Abate), which was usually dropped from helicopters or small aircraft. Because of the appearance of temephos resistance in 1980 in some populations and species of the *S. damnosum* complex, some rivers were treated with other insecticides or with *Bacillus thuringiensis* subsp. *israelensis* (=*B. thuringiensis* serotype H-14). To hinder the spread of further resistance a rotation of different insecticides was initiated in 1982. Larviciding must continue in parts of the four countries added in 1986 until the year 2002. Since 1988 the OCP has been undertaking large-scale distribution of the microfilaricide ivermectin (Mectizan), which is given orally once or twice a year.

Results from the OCP have been most spectacular, and transmission of river blindness has ceased over most of the OCP area. In 1995 the African Programme for Onchocerciasis Control (APOC) was created to cover populations at risk in 19 countries outside the OCP. The objective is to establish, within 12 years, a sustainable community-based ivermectin treatment regimen, backed up with focal larviciding. Because ivermectin does not kill adult worms control needs to continue for 20 years so that the reservoir of infection in the human population (adult onchocercal worms) dies out.

FURTHER READING

Anon. (1998) Mectizan and onchocerciasis; a decade of accomplishment and prospects for the future; the evolution of a drug into a development concept. *Annals of Tropical Medicine and Parasitology*, **92** (Suppl. 1), 179pp.

Crosskey, R.W. (1990) *The Natural History of Blackflies*. Chichester: Wiley.

Davies, J.B. (1994) Sixty years of onchocerciasis vector control: a chronological summary with comments on eradication, reinvasion, and insecticide resistance. *Annual Review of Entomology*, **39**, 23–45.

Davies, J.B. and Crosskey, R.W. (1991) *Simulium* – vectors of onchocerciasis. WHO/VBC/91.992. World Health Organization mimeographed document.

De Villiers, P.C. (1987) *Simulium* dermatitis in man – clinical and biological features in South Africa. *South African Medical Journal*, **71**, 523–5.

Hougard, J.-M., Yaméogo, L., Sékétéli, A., Boatin, B. and Dadzie, K.Y. (1997) Twenty-two years of blackfly control in the onchocerciasis control programme in West Africa. *Parasitology Today*, **13**, 425–8.

Kim, K.C. and Merritt, R.W. (eds.) (1988) *Black Flies: Ecology, Population Management, and Annotated World List 1987*. University Park and London: Pennsylvania State University. (Book incorrectly dated 1987, but published 1 March 1988.)

Molyneux, D.H. and Davies, J.B. (1997) Onchocerciasis control: moving towards the millennium. *Parasitology Today*, **13**, 418–25.

Raybould, J.N. and White, G.B. (1979) The distribution, bionomics and control of onchocerciasis vectors (Diptera: Simuliidae) in eastern Africa and the Yemen. *Tropenmedizin und Parasitologie*, **30**, 505–47.

Service, M.W. (1977) Methods for sampling adult Simuliidae, with special reference to the *Simulium damnosum* complex. *Tropical Pest Bulletin*, **5**, 1–48.

World Health Organization (1995) Onchocerciasis and its control. Report of a WHO expert committee on onchocerciasis control. *World Health Organization Technical Report Series*, **852**, 1–103.

5

Phlebotomine sandflies (Phlebotominae)

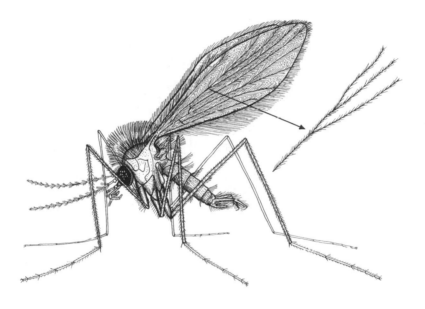

There are some 700 species of phlebotomine sandflies in six genera within the subfamily Phlebotominae of the family Psychodidae. Species in three genera – *Phlebotomus*, *Lutzomyia* and *Sergentomyia* – suck blood from vertebrates; the former two are the more important medically as they contain disease vectors. The genus *Phlebotomus* occurs only in the Old World, especially in southern parts of the northern temperate areas such as the Mediterranean region. The genus also occurs in the Old World tropics, but there are not many species in tropical Africa, especially West Africa. Most *Phlebotomus* species inhabit semiarid and savannah areas in preference to forests. *Lutzomyia* species by contrast are found only in the New World tropics, occurring mostly in the forested areas of Central and South America.

Sergentomyia species are also confined to the Old World, being especially common in the Indian subregion, but also occurring in other areas such as Africa and Central Asia. A few species of *Sergentomyia* bite people, but they are not disease vectors.

Adult flies are often called sandflies. However, this can be confusing because in some parts of the world the small biting midges of the Ceratopogonidae (Chapter 6) and blackflies (Simuliidae, Chapter 4) are called sandflies. The most important species include *Phlebotomus papatasi*, *P. sergenti*, *P. argentipes*, *P. ariasi*, *P. perniciosus* and species in the *Lutzomyia longipalpis* and *Lutzomyia flaviscutellata* complexes. In both the Old and New Worlds sandflies are vectors of leishmaniasis, the virus responsible for sandfly fever, and an organism causing a disease called bartonellosis (Carrión's disease).

5.1   EXTERNAL MORPHOLOGY

No details are given for distinguishing between the adults of *Phlebotomus*, *Lutzomyia* and *Sergentomyia* because this requires specialized knowledge and detailed examination; generally species biting people in the Old World will be *Phlebotomus* and in the New World *Lutzomyia*.

Adult phlebotomine sandflies can be readily recognized by their minute size (1.3–3.5 mm in length), hairy appearance, relatively large black eyes and their relatively long and stilt-like legs. The only other blood-sucking flies which are as small as this are some species of biting midges (Ceratopogonidae), but these have non-hairy wings and differ in many other details (Chapter 6). Phlebotomine sandflies have the head, thorax, wings and abdomen densely covered with long hairs. The antennae are long and composed of small bead-like segments with short hairs; they are similar in both sexes. The mouthparts are short and inconspicuous and adapted for blood-sucking. At their base are a pair of five-segmented maxillary palps which are relatively conspicuous and droop downwards.

Wings are lanceolate in outline and quite distinct from the wings of

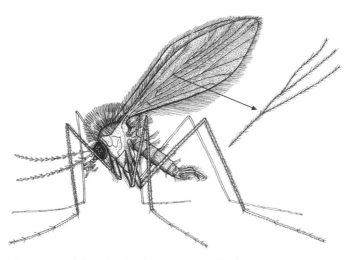

Figure 5.1 Adult male phlebotomine sandfly showing terminal genital claspers, and a diagrammatic representation of double branching of wing vein 2.

other biting flies. The Phlebotominae can be distinguished from other sub-families of the family Psychodidae, which they may superficially resemble, by the wings. In sandflies the wings are held erect over the body when the fly is at rest, whereas in non-biting psychodid flies they are folded, roof-like, over the body. The wing venation also differs. In phlebotomine sand-flies, but not in the other subfamilies of Psychodidae, vein 2 branches twice, although this may not be apparent unless most of the hairs are rubbed from the wing veins (Fig. 5.1).

The abdomen is moderately long and in the female more or less rounded at the tip. In males it terminates in a prominent pair of genital claspers which give the end of the abdomen an upturned appearance.

Identification of adult phlebotomine sandflies to species is difficult and usually necessitates the examination of internal structures, such as the arrangement of the teeth on the cibarial armature, the shape of the sper-matheca in females, and in males the structure of the external genitalia (ter-minalia).

5.2   LIFE CYCLE

The minute eggs (0.3–0.4 mm) are more or less ovoid in shape and usually brown or black, and careful examination under a microscope reveals that they are patterned, as shown in Fig. 5.2. Some 30–70 eggs are laid singly at each oviposition. They are deposited in small cracks and holes in the ground, at the base of termite mounds, in cracks in masonry, on stable floors, in poultry houses, amongst leaf litter and in between buttress roots

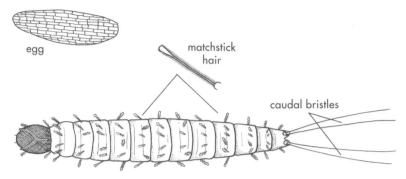

Figure 5.2 Egg of a phlebotomine sandfly showing mosaic-type pattern, and a larva with matchstick hairs and caudal bristles.

of forest trees, etc. The type of oviposition site varies greatly according to species.

Although eggs are not laid in water they require a moist microhabitat with a high humidity. They are unable to withstand desiccation and hatch after about 6–17 days under optimum conditions. However, hatching may be prolonged in cooler weather. Larvae are mainly scavengers, feeding on organic matter such as fungi, decaying forest leaves, semi-rotting vegetation, animal faeces and decomposing bodies of arthropods. Although some species, especially of the genus *Phlebotomus*, occur in semiarid areas, the actual larval habitats must have a high degree of humidity. Larvae can usually survive if their breeding places are temporarily flooded by migrating to drier areas.

There are four larval instars. The mature larva is 3–6 mm long and has a well-defined black head which is provided with a pair of small mandibles; the body is greyish or yellowish and 12-segmented (Fig. 5.2). The first seven abdominal segments have small pseudopods but the most striking feature of the larva is the presence, on the head and all body segments, of conspicuous thick bristles with feathered stems, which in many species have slightly enlarged tips. They are called matchstick hairs and identify the larvae as those of phlebotomine sandflies. In most species the last abdominal segment bears two pairs of conspicuous long hairs called the caudal setae. The first-instar larva has two single bristles not two pairs.

Larval development is completed after 19–60 days, the duration depending on species, temperature and availability of food. In temperate areas and arid regions species may overwinter as diapausing fully grown larvae. Prior to pupation the larva assumes an almost erect position in the habitat, the skin then splits open and the pupa wriggles out. The larval skin is not completely cast off but remains attached to the end of the pupa. The

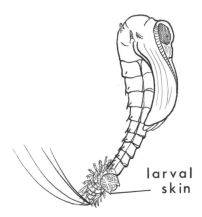

larval
skin

Figure 5.3 Pupa of a phlebotomine sandfly with larval skin still attached.

presence of this skin, with its characteristic two pairs of caudal bristles, aids in the recognition of the phlebotomine pupae. The pupal shape is as shown in Fig. 5.3. Adults emerge from the pupae after about 5–10 days. The life cycle, from oviposition to adult emergence, may be 30–60 days, but extends to several months in some species with diapausing larvae. For instance, in temperate areas adults die off in late summer or autumn and species overwinter as larvae, with the adults emerging the following spring. Apart from this the duration of the life cycle may be longer during cooler periods of the year.

It is usually extremely difficult to find larvae or pupae of sandflies and relatively little is known about their biology and ecology.

5.2.1 Adult behaviour
Both sexes feed on plant juices and sugary secretions but females in addition suck blood from a variety of vertebrates, including livestock, dogs, urban and wild rodents, snakes, lizards and amphibians; a few species feed on birds. In the Old World many *Phlebotomus* species bite people whereas most species of *Sergentomyia* feed mainly on reptiles and rarely bite humans. In the tropical Americas *Lutzomyia* species feed on a variety of mammals including humans. Biting is usually restricted to crepuscular and nocturnal periods but people may be bitten during the day in darkened rooms, or in forests during overcast days. Most species feed out of doors (exophagic) but some species also feed indoors (endophagic). A few species are autogenous.

Adults are weak fliers and do not usually disperse more than a few hundred metres from their breeding places. Consequently biting may be localized to a few areas. However, it is known that adults of at least some species can fly up to 2.2 km over a few days. Sandflies have a characteristic hopping type of flight so that there may be several short flights and landings before females settle on their hosts. Windy weather inhibits flight activities and biting. Because of their very short mouthparts sandflies are unable to bite through clothing.

During the day adult sandflies rest in sheltered, dark and humid sites, but on dry surfaces, such as on tree trunks, on ground litter and foliage of forests, in animal burrows, termite mounds, tree-holes, rock fissures, caves, cracks in the ground and inside human and animal habitations. Species that commonly rest in houses (endophilic) before or after feeding on humans are often referred to as domestic or peridomestic species. Examples are *Phlebotomus papatasi* in the Mediterranean area and the *Lutzomyia longipalpis* complex in South America.

In temperate areas of the Old World sandflies are seasonal in their appearance and adults occur only in the summer months. In tropical areas some species appear to be common more or less throughout the year, but in other species there may be well marked changes in abundance of adults related to the dry and wet seasons.

5.3   MEDICAL IMPORTANCE

5.3.1   Annoyance

Apart from their importance as disease vectors, sandflies may constitute a serious, but usually localized, biting nuisance. In previously sensitized people their bites may result in severe and almost intolerable irritations, a condition known in the Middle East as harara.

5.3.2   Leishmaniasis

Leishmaniasis is a term used to describe a number of closely related diseases caused by several distinct species, subspecies and strains of *Leishmania* parasites. The diseases occur in three main clinical forms: cutaneous, mucocutaneous and visceral leishmaniasis. A fourth, less common form is termed diffuse cutaneous leishmaniasis. Phlebotomine sandflies are the only vectors. The epidemiology of leishmaniasis is complex, involving not only different parasite species but strains of parasites, and only a simplified account of the disease and its vectors can be presented here.

Basically parasites (amastigotes) are ingested by female sandflies with a blood-meal and multiply in the gut. They develop a flagellum and attach themselves to either the mid-gut or hind-gut wall. After further development they become infective metacyclic promastigotes and migrate to the anterior part of the mid-gut and from there to the oesophagus. After 4–12 days from taking an infective blood-meal the metacyclic forms may be found in the mouthparts, from which they are introduced into a new host during feeding. Infective flies often probe more often than uninfected flies, thus maximizing transmission of parasites during blood-feeding. Previous feeding by females on sugary substances, mostly obtained from plants, is essential for not only the survival of the sandfly but the development of the parasites to the infective form.

Most types of leishmaniasis are zoonoses. The degree of involvement of humans varies greatly from area to area. The epidemiology is largely determined by the species of sandflies, their ecology and behaviour, the availability of a wide range of hosts and also by the species and strains of *Leishmania* parasites. In some areas, for example, sandflies will transmit the disease almost entirely among wild or domesticated animals, with little or no human involvement, whereas elsewhere animals may provide an important reservoir of infection for humans. In India the disease may be transmitted between people by sandflies, with animals taking no identifiable part in its transmission.

Cutaneous leishmaniasis in the Old World, known also as Oriental sore, occurs mainly in arid areas of the Middle East to northwestern India and central Asia, in North Africa and various areas in East, West and southern Africa. In the New World it is found mainly in forests from Mexico to northern Argentina. Several species of parasites are involved, including *Leishmania major* and *Le. tropica* in the Old World, and species in the *Le. braziliensis* and *Le. mexicana* complexes in the Americas. *Leishmania major* is usually zoonotic and in most of its range gerbils are the reservoir hosts; *Le. tropica* occurs in densely populated areas and, although dogs are also infected, humans are the main reservoir host. Important vectors of cutaneous leishmaniasis include *Phlebotomus papatasi* (*Le. major*), *P. longipes* (*Le. aethiopica*) and *P. sergenti* (*Le. tropica*).

Mucocutaneous leishmaniasis (espundia) is a severely disfiguring disease found from Mexico to Argentina. It is mainly caused by *Leishmania braziliensis* and *Le. panamensis*, parasites which normally infect a wide variety of forest rodents, marsupials, armadillos, edentates, primates, sloths and also domestic dogs. The disease is spread by several species of *Lutzomyia* including *L. wellcomei*, *L. intermedia* and *L. amazonensis*.

Diffuse cutaneous leishmaniasis is a form that causes widespread cutaneous nodules or macules over the body and usually results from infection with *Leishmania aethiopica* in Ethiopia and Kenya or with *Le. amazonensis* in South America.

Visceral leishmaniasis, often referred to as kala-azar, is caused by *Leishmania donovani* in most areas of its distribution, such as India, Bangladesh, Sudan, East Africa, and Ethiopia. Among the vectors are *P. argentipes* and *P. orientalis*. In Sudan rodents have been found infected but most transmission seems to be from person to person. In the Mediterranean basin, Iran, and central Asia, including northern and central China, *Leishmania infantum* is the parasite, and the vectors include *P. ariasi*, *P. perniciosus* and *P. chinensis*. Dogs and foxes are the most important reservoirs. Visceral leishmaniasis also occurs sporadically in Central and South America, where the parasite is *Leishmania infantum* (=*Le. chagasi* of some authors), and is transmitted by species in the *Lutzomyia longipalpis* complex. Wild foxes and domesticated dogs are reservoirs.

5.3.3   Bartonellosis
Bartonellosis is sometimes called Oroya fever or Carrión's disease and is encountered in arid mountainous areas of the Andes in Peru, Ecuador and Colombia. It is caused by a small rod-like micro-organism named *Bartonella bacilliformis* and is transmitted by *Lutzomyia verrucarum* and possibly by other *Lutzomyia* species. Transmission is probably entirely by contamination of the mouthparts.

5.3.4   Sandfly fever
Sandflies can transmit several viral diseases, but the most important is sandfly fever, sometimes called papataci fever or 3 day fever, of which there are two distinct virus serotypes. Females become infective 6–10 days after taking an infected blood-meal. Infected females can lay eggs containing the virus and these eventually give rise to infected adults. This is an example of trans-ovarial transmission, a phenomenon that is more common in the transmission of various tick-borne diseases (Chapter 17). Sandfly fever occurs mainly in the Mediterranean region, but also extends up the Nile into Egypt and the Middle East; the most important vector is *P. papatasi*. The disease also occurs in India, Pakistan and probably China, and in some of these areas phlebotomines other than *P. papatasi* are the vectors.

5.4   CONTROL
Leishmaniasis has in the past not usually been considered a sufficiently important disease to justify much expenditure on controlling the vectors, and consequently there have been relatively few campaigns devoted to sandfly control. Phlebotomine sandflies are, however, very susceptible to insecticides; the only reported case of resistance is *P. papatasi* to DDT in northeastern India. In nearly all areas where residual insecticides, particularly DDT, have been used to control *Anopheles* vectors of malaria there have apparently been dramatic reductions in sandfly populations followed by interruption of leishmaniasis transmission. With the cessation of spraying, sandflies usually return, and transmission is renewed. In addition to DDT, HCH, malathion, fenitrothion (Simuthion) and propoxur (Baygon), the synthetic pyrethroids, have been used to try to control sand-

flies. Where sandflies bite and rest out of doors such control strategies are not feasible.

If the outdoor resting sites are known (e.g. animal shelters, stone walls, tree trunks, termite hills) these can be sprayed with residual insecticides. Insecticidal fogging of outdoor resting sites is used in Jordan for the control of *P. papatasi*, although this approach usually gives only temporary control.

Personal protection can be achieved by applying efficient insect repellents such as diethyltoluamide (deet), dimethylphthalate (dimp) or trimethyl pentanediol. Mesh screens or nets with very small holes can afford some protection, but a disadvantage is that they reduce ventilation, causing it to be unpleasantly hot in screened houses or under sandfly bed-nets. Such nets, or even mosquito nets with larger holes, can be impregnated with pyrethroids such as permethrin and deltamethrin to give protection against biting for up to 6 months, so long as the nets are not washed (see p. 49 for more details on the use of insecticide-impregnated bed-nets for protection against malaria vectors).

Control of sandfly larvae remains impossible because the breeding sites of most species are unknown, and even when they can be identified it is usually impractical to do anything to reduce larval numbers.

Because most leishmaniasis transmission involves reservoir animals, such as rodents and dogs, in some areas control focuses on killing such hosts. For example in some areas of North Africa clearing vegetation around villages reduces gerbil populations, and in parts of Brazil, China and the Mediterranean area infected dogs and other reservoir hosts are destroyed.

FURTHER READING

Hart, D.T. (ed.) (1989) *Leishmaniasis: The Current Status and New Strategies for Control*. New York and London: Plenum Press.

Hide, G., Mottram, J.C., Coombs, G.H. and Holmes, P.H. (1996) *Trypanosomiasis and Leishmaniasis: Biology and Control*. Wallingford: CAB International.

Killick-Kendrick, R. (1990) Phlebotomine vectors of the leishmaniases: a review. *Medical and Veterinary Entomolology*, **4**, 1–24.

Killick–Kendrick, R. (1999) The biology of phlebotomine sand flies. *Clinics in Dermatology*, **17**, 279–89.

Lainson, R. (1983) The American leishmaniases: some observations on their ecology and epidemiology. *Transactions of the Royal Society of Tropical Medicine and Hygiene*, **77**, 569–96.

Lainson, R. (1989) Demographic changes and their influence on the epidemiology of the American leishmaniases. In *Demography and Vector-Borne Diseases*, ed. M.W. Service, pp. 85–106. Boca Raton, Florida: CRC Press.

Lane, R.P. (1991) The contribution of sandfly control to leishmaniasis control. *Annales de la Société belge de Médicine Tropicale*, **71** (Suppl.), 65–74.

Peters, W. and Killick-Kendrick, R. (eds.) (1987) *The Leishmaniases in Biology and*

Medicine; vol. 1, *Biology and Epidemiology*, vol. 2, *Clinical Aspects and Control*. London: Academic Press.

Walton, B.C., Wijeyaratne, P. and Moddaber, F. (eds.) (1988) *Research on Control Strategies for the Leishmaniases*. Proceedings of an international workshop held in Ottawa, Canada, 1–4 June 1987. Ottawa: IDRC.

Ward, R.D. (1990) Some aspects of the biology of phlebotomine sandfly vectors. *Advances in Disease Vector Research*, **6**, 91–126. (Reprints and chapters incorrectly dated 1989.)

World Health Organization (1990) Control of the leishmaniases. Report of a WHO Expert Committee. *World Health Organization Technical Series*, **793**, 1–158.

6

Biting midges (Ceratopogonidae)

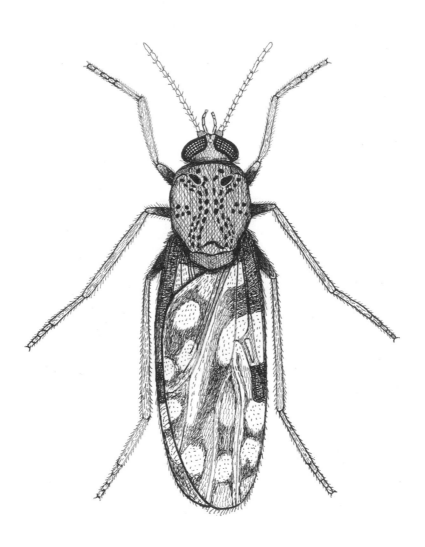

Biting midges belong to the family Ceratopogonidae, of which there are some 5000 species in over 120 genera. Midges have a more or less world-wide distribution, the most widely distributed genus being *Culicoides* which is found from tropical to subarctic areas. Only species in the subgenus *Lasiohelea* of the genus *Forcipomyia* and in the genera *Leptoconops*, *Austroconops* and *Culicoides* feed on vertebrates including humans. Of these the most important are the *Culicoides*, of which over 1200 species have been described. In many parts of the world species of *Culicoides*, and in the Americas also *Leptoconops*, can constitute a serious biting problem. In Africa species of the *Culicoides milnei* group, *C. grahamii* and *C. inornatipennis* are vectors of the filarial worms *Mansonella perstans*, while *C. grahamii* and possibly the *C. milnei* group are vectors of *Mansonella streptocerca*. *Culicoides furens* is a vector of *Mansonella ozzardi* in the Americas. These parasites are usually regarded as non-pathogenic to humans.

6.1   EXTERNAL MORPHOLOGY

A generalized description is given of the adults of *Culicoides*. Adult *Culicoides* are sometimes known as midges or biting midges, and, especially in the Americas, as 'no-see-ums' or 'punkies', and in Australia and some other countries as sandflies. This last name is unfortunate and should be avoided because phlebotomines (Chapter 5) and occasionally simuliids (Chapter 4) may also be referred to as sandflies. The most appropriate common name is biting midges; this terminology serves to distinguish them from other small non-biting flies which are often referred to as midges.

The adults are very small insects only 1–2 mm long, and with the phlebotomines constitute the smallest biting flies attacking humans.

The head bears a prominent pair of eyes, a pair of short five-segmented palps and a pair of relatively long filamentous antennae. As in mosquitoes, males do not take blood-meals and have feathery or plumose antennae, whereas the blood-sucking females have non-plumose antennae. The biting mouthparts are very small and inconspicuous and they do not project forwards but hang down vertically from the head. The arrangement and structure of the mouthparts are very similar to those of simuliids (Chapter 4). In some species the thorax is covered dorsally with very small but distinct black spots and markings. In addition to these dark markings, a pair of shiny black small but elongated depressions, known as the humeral pits, are present in all *Culicoides* species on the dorsal surface of the anterior part of the thorax. The presence of these pits distinguishes the genus *Culicoides* from *Lasiohelea* and *Leptoconops*. The wings are short and relatively broad, and apart from the first veins their venation is faint. The wings lack scales but in many species are covered with minute hairs – these are seen only under a dissecting microscope. In most *Culicoides* species the wings have contrasting dark and milky white spots or patches (Fig. 6.1). In

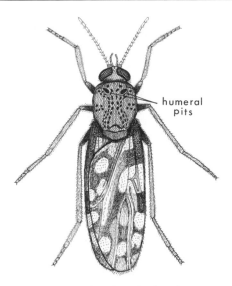

Figure 6.1 Adult *Culicoides* showing patterned wings and thoracic humeral pits.

life the wings are placed over the abdomen like the blades of a closed pair of scissors (Fig. 6.1). The legs are relatively short.

The abdomen is dull grey, yellowish brown or blackish and in the female is more or less rounded at its tip; in the male there is a pair of small, but conspicuous, claspers.

6.2   LIFE CYCLE

The eggs are dark, cylindrical or curved and banana-shaped, and are about 0.5 mm long (Fig. 6.2c). They are laid in batches of about 30–250 on the surface of mud, wet soil, especially that near swamps and marshes including salt-water marshes, or decaying leaf litter, humus, manure, or on plants and other objects near, or partially submerged in, water, in tree-holes, in semi-rotting vegetation and in the cut stumps of banana plants (for example *Culicoides milnei* group and *C. grahamii*). The type of oviposition site selected depends on the species.

Eggs usually hatch within about 2–9 days, depending on temperature and species; some temperate species overwinter as eggs. There are four larval instars and a fully grown larva is cylindrical, whitish, about 5–6 mm long and nematode-like. It has a small, dark (but unpigmented in *Leptoconops*) conical-shaped head followed by 12 body segments (several books incorrectly say 11 segments). The last segment terminates in two four-lobed retractile papillae (Fig. 6.2*a*). These are not always readily seen in preserved larvae because they are often retracted within the last abdominal segment (Fig. 6.2*b*). *Culicoides* larvae are best recognized by the

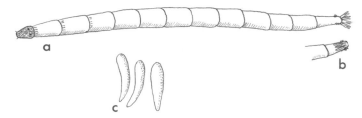

Figure 6.2 (a) Larva of a *Culicoides* with two four-lobed retractile papillae extending beyond the last abdominal segment; (b) last abdominal segment with papillae retracted; (c) *Culicoides* eggs.

combination of a small dark head followed by a segmented body devoid of any obvious structures and, when they are extruded, by the presence of terminal papillae. When alive they can also be recognized by their serpentine swimming motions.

Larvae feed mainly on decaying vegetable matter and occur in many different types of habitats, including fresh- or salt-water marshes and swamps, edges of ponds, boggy and semi-waterlogged areas, and in specialized habitats such as horse and cow excreta, tree-holes and rotting cacti. When the water level in swamps and marshes rises the larvae of many species migrate to the damp soil and mud at the edges to avoid becoming completely submerged. Some important pest species breed in sandy areas near the seashore. Larvae are difficult to find and are rarely encountered unless special surveys are made to collect them.

In warm countries larval development is completed within 14–25 days, but in temperate regions many species overwinter and remain as larvae for 7 months or more. Species occupying marshy habitats frequently migrate to the drier peripheral areas for pupation. However, in species that are aquatic the pupae float at the water surface.

The pupa (Fig. 6.3) is 2–4 mm long and readily recognized by the following combination of characters: (i) a pair of breathing trumpets on the cephalothorax which appear to be composed of two segments; (ii) abdominal segments bearing small but conspicuous tubercles ending in a fine hair; and (iii) a prominent pair of horn-like processes on the last abdominal segment. The pupal period lasts 3–10 days.

6.2.1   Adult behaviour

Adults of both sexes feed on naturally occurring sugar solutions. In addition females take blood-meals from humans and a variety of mammals and birds. Adults bite at any time of the day or night, but many species are particularly active and troublesome in the evenings and first half of the night. In contrast *C. grahamii* bites mainly in the early part of the mornings.

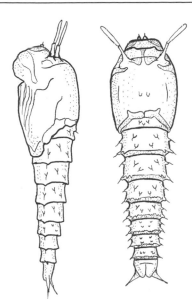

Figure 6.3 Lateral and dorsal views of a *Culicoides* pupa.

Because of their short mouthparts, biting midges are not as successful in biting through clothing as are mosquitoes, tabanids and tsetse-flies, all of which have a longer proboscis. For this reason midges often appear in swarms or clouds around the head, biting the face, especially the forehead and scalp. They also bite other exposed parts such as the hands and arms, and the legs of people wearing shorts. Most species bite only outdoors, but a few including the *C. milnei* group and *C. grahamii* will enter houses to feed on people (endophagic). About 30 species are autogenous, that is they can mature and lay the first batch of eggs without a blood-meal. Adults normally fly only a few hundred metres from their larval habitats, but they may be dispersed considerably further by the wind. However, some species of both *Culicoides* and *Leptoconops* are known to fly 2–3 km without the assistance of wind.

6.3   MEDICAL IMPORTANCE

6.3.1   Annoyance

Biting midges are very small, but what they lack in size they can make up for in numbers – as has been said, one midge is an entomological curiosity, a thousand sheer hell! In several areas of the world, as dissimilar as the west coast of Scotland, the Caribbean and the sunny regions of California and Florida, biting midges can be a serious economic threat to the tourist industry. The persistent biting of large numbers of midges can make outdoor recreational activities impossible, not only at dusk but often during much of the day. In some areas they have even prevented the continuation of harvesting and other outdoor work during the evenings.

Important pest species in southern areas of North America down to Brazil include *C. furens* which breeds in salt marshes and other saline coastal habitats. In North America *Leptoconops torrens* and *L. bequaerti*, which also breed in sandy soils and coastal areas, can be serious pests. In Europe *C. impunctatus* and many other species are troublesome biters, while in Madagascar, the Seychelles and Brunei *L. spinosifrons* can cause a considerable biting problem.

6.3.2   Filarial infections

A few *Culicoides* species are vectors of parasites to humans. In Africa, especially West and Central, but also parts of East Africa as far south as Zimbabwe, *Mansonella perstans* is transmitted to humans by the *C. milnei* group and possibly also by *C. grahamii*. *Mansonella perstans* also occurs in Trinidad and South America where it is transmitted by other *Culicoides* species. In the rain forests of Ghana, Burkina Faso, Nigeria, Cameroon and Zaire *Mansonella streptocerca* is transmitted by *C. grahamii*, and possibly by the *C. milnei* group. These species breed in the rotting cut stumps of banana and plantain plants.

In Mexico, Panama, the West Indies and South America *Mansonella ozzardi* is transmitted by *Culicoides* species, mainly *C. furens*, but other species such as *C. phlebotomus* are also likely to be involved. (The role of simuliids in the transmission of this filarial parasite is discussed in Chapter 4.)

Microfilariae of all these parasites are non-periodic, they are ingested with a blood-meal and pass through a similar developmental cycle to other filarial parasites in mosquitoes. That is, they undergo morphological changes, invade the thoracic flight muscles, moult twice and then migrate to the head and after about 9–12 days pass down the proboscis. The infective third-stage larvae are deposited on the skin of the host when the female takes a blood-meal. The salivary glands of *Culicoides* play no part in the transmission of these parasites. None of the three filarial parasites carried by midges appears to cause much harm and they are usually regarded as non-pathogenic, although morbidity or allergic reactions may sometimes occur.

6.3.3   Arboviruses

The only known arbovirus transmitted to humans by midges is Oropouche virus, a relatively unimportant infection which occurs in Brazil, Trinidad and Colombia. In marked contrast *Culicoides* transmit several arboviruses to animals, including bluetongue virus which causes severe disease in sheep and African horse sickness which can be fatal to horses.

In general the Ceratopogonidae are not considered very important vectors of disease to humans, whereas they can be of considerable veterinary importance.

6.4 CONTROL

Because many of the major pest species breed in extensive and often diffuse habitats, such as fresh- and salt-water marshes and wet coastal sands, whose limits are usually difficult to define, it is usually very difficult to reduce larval breeding substantially. For effective control often large areas of land or marsh must be drained or sprayed with insecticides.

Although larval habitats can be eradicated by draining them or filling them in, this is often costly, laborious and in many areas impractical. Sometimes semi-aquatic sites such as muddy or marshy areas, can be impounded and flooded under 5–8 cm of water to ensure that the soil is never exposed, thus destroying suitable habitats. This type of environmental control can be effective against some species but ineffective against others; it can also be very expensive. If maintained such methods have the advantage of giving permanent control. However, although they avoid contaminating the environment with insecticides they in themselves result in completely changing the habitat.

Insecticidal spraying of larval breeding sites has sometimes been effective, but heavy rainfall is required to wash the insecticide through the surface vegetation to the underlying soil and mud harbouring the midge larvae. Although in the past insecticides such as dieldrin, diazinon and chlorpyrifos (Dursban) have been used, greater public awareness of environmental contamination would these days in many cases prevent such indiscriminate insecticidal applications.

Thermal insecticidal aerosols or ultra-low-volume (ULV) applications have sometimes been used to kill adults resting in vegetation, but the effects are very short-lived and sprayed areas are soon invaded by midges flying in from unsprayed areas.

Limited personal protection can be achieved by the use of suitable repellents such as diethyltoluamide (deet), dimethylphthalate (dimp) or trimethyl pentanediol. Mosquito nets and screening used to keep out house-flies and mosquitoes may not exclude the much smaller biting midges. To prevent them from passing through protective nets and screens a very small mesh size must be used, but a disadvantage is that this substantially reduces air flow and ventilation. Nets and screens can be treated with pyrethroids such as permethrin and deltamethrin. Such treated nets may remain effective for 6 months or more, after which they need to be re-impregnated.

FURTHER READING

Holbrook, F.R. (1985) An overview of *Culicoides* control. In *Bluetongue and Related Orbiviruses*, ed. T.L. Barber and M.M. Jochim, pp. 607–9. Progress in Clinical and Biological Research **178**. New York: Alan J. Liss.

Kettle, D.S. (1965) Biting ceratopogonids as vectors of human and animal diseases. *Acta Tropica* **22**, 356–62.

Kettle, D.S. (1969) The ecology and control of blood-sucking ceratopogonids. *Acta Tropica*, **26**, 235–48.

Kettle, D.S. (1977) Biology and bionomics of blood-sucking ceratopogonids. *Annual Review of Entomology*, **22**, 33–51.

Linley, J.R. and Davies, J. B. (1971) Sandflies and tourism in Florida and the Bahamas and Caribbean area. *Journal of Economic Entomology*, **64**, 264–78.

Linley, J.R., Hoch, A. L. and Pinheiro, F.P. (1983) Biting midges (Diptera: Ceratopogonidae) and human health. *Journal of Medical Entomology*, **20**, 347–64.

Mellor, P., Boorman, J.P.T. and Baylis, M. (2000) *Culicoides* biting midges. *Annual Review of Entomology*, **45** (in press).

Nathan, M. B. (1979) The prevalence and distribution of *Mansonella ozzardi* in coastal north Trinidad, W.I. *Transactions of the Royal Society of Tropical Medicine and Hygiene*, **73**, 299–302.

Reynolds, D.G. and Vidot, A. (1978) Chemical control of *Leptoconops spinosifrons* in the Seychelles. *Pest Articles and News Summaries*, **24**, 19–26.

7

Horseflies (Tabanidae)

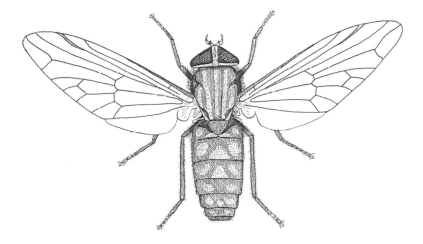

The tabanids are large biting flies generally called horseflies, although other vernacular names include gadflies, stouts, greenheads (some species of *Tabanus*), clegs (*Haematopota*) and deer-flies (*Chrysops*). They all belong to the family Tabanidae, which comprises some 4000 species in 30 genera. The most important species medically belong to the genera *Tabanus*, *Chrysops* and *Haematopota*. Tabanids have been incriminated in the spread of tula-raemia in North America and the former USSR, and anthrax, and it has been suggested they might be involved in the transmission of Lyme disease (usually transmitted by hard ticks). But the main medical impor-tance of the Tabanidae is that species of *Chrysops*, mainly *C. silaceus* and *C. dimidiatus*, are vectors in West and Central Africa of the filarial worm *Loa loa*.

Tabanids possibly play a very minor role in the mechanical transmission of human and animal trypanosomiasis. In Central and South America tabanids are involved in the transmission of a form of *Trypanosoma vivax* to cattle and sheep.

The Tabanidae have a world-wide distribution. Species of *Tabanus* and *Chrysops* are found in temperate and tropical areas, but *Haematopota* is absent from South America and Australasia and is uncommon in North America.

7.1   EXTERNAL MORHPOLOGY

A generalized description is presented of the Tabanidae, with special refer-ence to the genera *Chrysops*, *Tabanus* and *Haematopota*.

Tabanids are medium to very large flies (6–30 mm long). Many, espe-cially of the genus *Tabanus*, are robust and heavily built, and this genus con-tains the largest biting flies, some attaining a wing span of 6.5 cm. The colouration of tabanids varies from very dark brown or black to lighter reddish-brown, yellow or greenish; frequently the abdomen and thorax have stripes or patches of contrasting colours (Fig. 7.1). The head is large and, viewed from above, is more or less semicircular in outline (Fig. 7.2); it is often described as semilunar. The head bears a conspicuous pair of com-pound eyes which in life may be marked with contrasting iridescent colours, such as greens and reds or even purplish hues, arranged in bands, zig-zags or spots. Adults can be sexed by examination of their eyes. In the female there is a distinct space on top of the head between the eyes; this is known as a dichoptic condition (Fig. 7.2). In females of some species this space between the eyes may be narrow, whereas in others, especially *Chrysops*, it is quite large. In the males the eyes are so large that they occupy almost all of the head and either touch each other on top of the head or are very narrowly separated, this being known as a holoptic condition (Fig. 7.2).

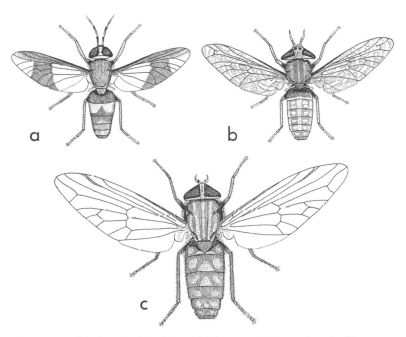

Figure 7.1 Adult female Tabanidae. (*a*) *Chrysops*; (*b*) *Haematopota*; (*c*) *Tabanus*.

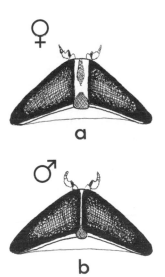

Figure 7.2 Dorsal view of tabanid heads. (*a*) Female with dichoptic eyes; (*b*) male with holoptic eyes.

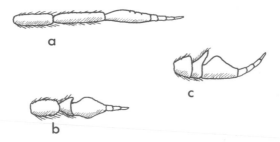

Figure 7.3 Antennae of adult tabanids. (*a*) *Chrysops*; (*b*) *Haematopota*; (*c*) *Tabanus*.

The antennae are relatively small but stout. They consist of three segments; the last is subdivided into three or four small divisions by annulations. Unlike the Muscidae, Glossinidae and Calliphoridae there is no antennal arista. The size and shape of the antennae serve to distinguish the genera *Chrysops*, *Haematopota* and *Tabanus* (Fig. 7.3). The mouthparts of female Tabanidae are stout and adapted for biting and, unlikely those of tsetse-flies, mosquitoes and *Stomoxys*, they do not project forwards but hang downwards from the head. Only female tabanids take blood-meals.

The thorax is stout and bears a pair of wings which have two submarginal and five posterior cells and a completely closed discal cell in approximately the centre of the wing (Fig. 7.4). Although the wing venation alone may not be sufficient to identify the Tabanidae from all other Diptera, it nevertheless serves as a useful guide when considered with other characters, such as the shape and structure of the antennae and biting mouthparts. The wings may be completely clear and devoid of colour or have areas of brown colouration, or they may be distinctly banded, or appear mottled or speckled due to the presence of greyish patches (Fig. 7.1). When adults are at rest the wings are placed either like a pair of open scissors over the abdomen or at a roof-like angle completely obscuring the abdomen. The presence or absence of coloured areas on the wings and the way in which they are held over the body provides useful additional characters for distinguishing between *Chrysops*, *Tabanus* and *Haematopota* (pp. 115–16).

The abdomen is usually broad and stout, and in unfed flies characteristically flattened dorsoventrally. It may be a more or less uniform dark

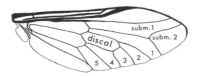

Figure 7.4 Wing of an adult tabanid showing discal cell, two submarginal (subm.) cells and five posterior cells (1–5).

brown, blackish, light brown, reddish-brown, yellowish or even greenish, or alternatively marked with contrasting coloured stripes or patches.

7.2   LIFE CYCLE

Males feed only on naturally occurring sugary secretions. Females also feed on sugary substances but in addition they bite a wide variety of mammals such as domestic animals, especially horses and cattle, deer and many other herbivores, carnivores, monkeys, reptiles and amphibians. A few species even attack birds. They also feed on people.

Some 100–1000 eggs, the number depending on the species, are deposited by female tabanids on the underside of objects such as leaves, grassy vegetation, plant stems, twigs, small branches, stones and rocks. These oviposition sites overhang, or are adjacent to, the larval habitats, which are mainly muddy, aquatic or semi-aquatic sites (p. 114). The eggs, which are firmly glued in an upright position in a large mass (up to 25 mm long) to the substrate, are covered with a coating that is impervious to water, i.e. they are waterproofed. They are usually arranged in a more or less lozenge-shaped pattern (Fig. 7.5a) and are mostly creamy white, greyish or blackish. Each egg measures 1–2.5 mm in length and is curved or approximately cigar-shaped. Eggs hatch after 4–14 days, the time depending on both temperature and species. After wriggling out of the eggs the young larvae drop down on to the underlying mud or water.

Larvae are cylindrical and rather pointed at both ends (Fig. 7.5c). They are creamy white, brown or even greenish but often with darkish pigmentation near the borders of the segments. There is a very small black head which can be retracted into the thorax. There are 11 well-differentiated

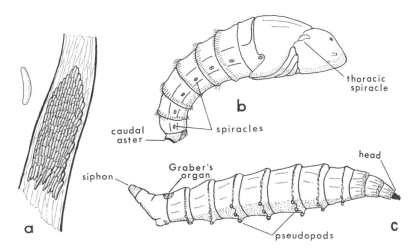

Figure 7.5 Immature stages of tabanids. (a) Single egg and egg mass glued to a piece of grass; (b) a pupa; (c) a larva.

body segments. Larvae are readily recognized by the prominent raised tyre-like rings which encircle most body segments. The first seven abdominal segments have one pair of lateral and two pairs of ventral (a total of six) conspicuous roundish protuberances called pseudopods. The presence of the prominent rings and these pseudopods readily identify larvae of tabanids. The last abdominal segment bears a short siphon dorsally which can be retracted into the abdomen, and a pyriform structure known as Graber's organ which is composed of 15 or fewer black globular bodies. This can readily be seen with the aid of a hand lens or microscope and is unique to tabanid larvae. This organ is well provided with nerves and seems to be sensory, but its exact function is not clearly understood.

Larvae live in mud, rotting vegetation, humus, damp soil, in shallow and often muddy water, at the edges of small pools, swamps, ditches or slowly flowing streams. In aquatic habitats larvae sometimes adhere to floating leaves, logs or other debris. A few species breed in tree-holes. Some species occur in brackish habitats, whereas larvae of certain *Tabanus* and *Haematopota* species are found in the relatively dry soils of pastures and in the earth near the bases of trees. Larvae breathe atmospheric oxygen through the short siphon at the posterior end of the body, and because they are poor swimmers aquatic species are usually found only in shallow waters. They move rather sluggishly in their muddy, aquatic or semi-aquatic environment. In some species, in particular *Chrysops*, the larvae are scavengers, feeding on detritus and a variety of dead and decaying vegetable and animal matter, whereas larvae of *Tabanus* and *Haematopota* are predacious or cannibalistic. Sometimes carnivorous tabanid larvae bite people, such as those working barefoot in ricefields. Such bites can be quite painful.

Larval development is prolonged. In both temperate and tropical countries many species spend 1–2 years as larvae, and several temperate species may remain as larvae for as long as 3 years. Larvae, however, are not easily found and relatively little is known about the life cycle of many species, but there appear to be 6–13 larval instars. Depending upon the species mature larvae may be 1–6 cm long. Prior to pupation mature larvae migrate to drier areas at the periphery of the larval habitat where they pupate. The pupa is partially buried in the mud or soil in an upright position and, superficially, looks like a chrysalis of a butterfly (Fig. 7.5*b*). It is 6–35 mm long, size depending on the species, distinctly curved and is usually brown. The head and thorax are combined to form a distinct cephalothorax which has a pair of lateral and relatively large, ear-shaped spiracles. The abdomen is composed of eight well-defined segments, the first seven of which are supplied with a pair of lateral spiracles, whereas segments 2–7 have an encircling row of small, backwardly directed spines. A few spines are also present ventrally on segment 8. The short terminal eighth abdominal segment is provided with six lobes which bear spine-like processes, known collectively as the caudal aster. The pupal period lasts about 5–20 days.

7.2.1   Adult behaviour

Females of most species feed during the daylight hours and are especially active in bright sunshine, but a few species are crepuscular and some feed at night. They locate their hosts mainly by sight (colour and movement), although olfactory stimuli such as carbon dioxide and other host odours also play a role in host location. Tabanids are powerful fliers and may disperse several kilometres.

Most tabanids inhabit woods and forests. Many *Chrysops* species are common in low-lying marshy scrub areas or swampy woods; some species, however, are found in more open savannah and grassland areas. Adults do not usually enter houses to feed, but *C. silaceus* is an exception. It is also attracted by smoke from wood burning and from forest fires. Other species may be found in houses, especially on windows, but often these have entered accidentally and are trying to escape.

Because of their large and rather broad mouthparts, bites from tabanids are deep and painful, sometimes especially so, and wounds inflicted by tabanids frequently continue to bleed after the female has departed. Due to their painful bites tabanids are frequently disturbed when feeding on either people or other hosts (e.g. horses, cattle, deer, monkeys). As a result several small blood-meals may be taken from the same or different hosts before the female has obtained a complete meal. This interrupted feeding behaviour increases their likelihood of being mechanical vectors of disease. Because of their preference for dark objects they often bite through coloured clothing when attacking Caucasians rather than biting exposed areas of pale skin; in this respect they behave like tsetse-flies.

In both temperate and tropical areas the occurrence of adults is seasonal. In temperate countries adults usually die off at the end of the summer and a new population emerges in the spring or summer of the following year. In the tropics the flies may not completely disappear in the dry months, but their numbers are normally much reduced. Maximum numbers of biting flies usually appear towards the beginning of the rainy season.

7.2.2   Identification of adult *Chrysops*, *Tabanus* and *Haematopota*

Chrysops species (deer-flies)

Chrysops species are medium-sized flies ranging in size from that of a house-fly to a tsetse-fly. In life most species have iridescent eyes, commonly with spots of red, green or purple. The wings, which are held partially over the abdomen in an open scissor-like fashion, usually have one or more transverse bands of brownish colour (Fig. 7.1*a*). In many species the abdomen is blackish with orange or yellow patches or bands.

The most reliable method of distinguishing *Chrysops* species from *Tabanus* and *Haematopota* is by the antennae. In *Chrysops* these are long; the second segment is long and cylindrical and has no projection (Fig. 7.3*a*), and the third segment has four small subdivisions. The hind tibiae have apical spurs; these are absent in *Tabanus* and *Haematopota*.

Chrysops has a world-wide distribution.

Tabanus species (horseflies, greenheads)

Tabanus species are medium to very large flies. The eyes are frequently brownish but may be iridescent; the markings are usually in the form of horizontal bands. The wings, which are held over the body much as in *Chrysops*, are often clear (Fig. 7.1*c*), but in some species there are dark markings.

Tabanus species are readily identified by the shape and size of the antennae. Both the second and third antennal segments have small, but distinct projections on the upper surface (Fig. 7.3*c*), the third segment has four small subdivisions and is usually distinctly curved upwards. The antennae are much shorter than those of *Chrysops* species, and are therefore less conspicuous.

Tabanus has a world-wide distribution.

Haematopota species (clegs, stouts)

Haematopota species are medium-sized dark grey flies which are easily distinguished from *Tabanus* and *Chrysops* by the fact that in life the wings are folded roof-like over the abdomen. Moreover, in nearly all species the wings are dusty grey and speckled or mottled (Fig. 7.1*b*). The eyes have zig-zag bands of iridescent colours. The antennae are similar to those of *Tabanus*, but usually a little longer. The third segment is straight, not curved as in *Tabanus*, has only three, not four, small subdivisions and does not bear a dorsal projection (Fig. 7.3*b*).

Haematopota species are not found in South America or Australasia, and only very few species occur in North America. They are common in Europe, Asia, Africa, India and the Far East.

7.3   MEDICAL IMPORTANCE

7.3.1   Minor infections

Because females tend to be intermittent feeders, and are often disturbed during feeding, tabanids are particularly liable to be mechanical vectors. In this way they can spread anthrax (*Bacillus anthracis*), and in North America and the former USSR they have been incriminated as spreading mechanically tularaemia (*Francisella tularensis*) from horses, rabbits and other rodents to humans. In North America *Chrysops discalis* has been identified as a vector of tularaemia, but other tabanid species are most likely involved. The disease is also commonly spread by handling infected rodents, by ixodid tick bites and by eating insufficiently cooked meat. It has also been suggested that Lyme disease (*Borrelia burgdorferi*), which is normally transmitted by ixodid ticks, may be transmitted by tabanids, but probably involving just mechanical transmission.

Tabanids are also mechanical vectors of *Trypanosoma evansi*, which causes a disease called Surra in camels, horses and dogs, often with fatal results; it does not infect people. In Central and South America tabanids are also mechanical vectors of *Trypanosoma vivax* to livestock. This form of *vivax* does not infect people.

Because of their painful bites, tabanids may sometimes constitute a pest nuisance and make outdoor activities, whether recreational or work, difficult. Some people develop severe allergic symptoms due to the large amount of saliva that is pumped into the wound to prevent blood clotting.

7.3.2 Loiasis

The only important and cyclical disease transmitted to humans by tabanids is loiasis, caused by the parasitic nematode *Loa loa* which undergoes a developmental cycle in the fly. This disease occurs principally in the equatorial rain forests of Sierra Leone westwards to Ghana and then from Nigeria across Central Africa, the southern Sudan and into western parts of Uganda. Diurnal periodic microfilariae are found in the peripheral blood of humans and are ingested by tabanids with their blood-meal. In *Chrysops* species, in particular *C. silaceus* and *C. dimidiatus*, many, but not all the microfilariae survive the process of blood digestion, penetrate the gut wall and migrate to the abdominal and thoracic fat bodies. Here they grow, moult twice and develop into shorter and fatter larval forms (2 mm long), which, after 10–12 days, are able to migrate down the proboscis. Infective larvae may be quite numerous in the fly and are commonly found in the abdomen and thorax as well as in the head and proboscis. When *Chrysops* feed on humans the infective worms migrate from other parts of the body to the proboscis and enter the skin. They eventually develop into adult worms which live in the subcutaneous tissues of people. Microfilariae of *Loa loa* are more or less absent from the blood circulation of people at night but appear in it during the day, especially in the morning, and are therefore readily picked up by *C. silaceus* and *C. dimidiatus* species which bite during the day. In areas of West and Central Africa where these flies are absent, other species, such as *C. distinctipennis* and *C. longicornis*, appear to be vectors. *Loa loa* from humans can be transmitted to monkeys, and some believe that loiasis may be a zoonosis. Whether or not this is true a similar parasite, previously considered to be *Loa loa* but now regarded as a closely related subspecies called *L. loa papionis*, occurs in some forest monkeys such as baboons. The microfilariae appear in the peripheral blood at night and are picked up by *C. centurionis* and *C. langi*, species which are mainly crepuscular and nocturnal and which feed in the tree canopy, where the monkeys are sleeping. It seems that the monkey form of *Loa* cannot be transmitted to humans.

7.4   CONTROL

There are very few practical control measures to combat tabanids. In theory efficient drainage not only to remove standing water but to dry out marshy and muddy areas might reduce the numbers breeding in these habitats, but the cost of both locating the larval habitats and carrying out drainage obviates this approach. Similarly, because of the difficulty of locating breeding places and their diffuse and large size, it is usually impossible to achieve control by the application of insecticides. Moreover, because the larvae of many species live below the surface of the ground, heavy dosage rates would be needed for the insecticide to penetrate through the surface soil and vegetation to reach the larvae; these problems are somewhat similar to those encountered in the control of ceratopogonid larvae (Chapter 6).

Some level of local control can sometimes be achieved by employing attractant traps to catch adults, such as coloured screens coated with adhesives. But there are no really effective control methods against tabanids. Some degree of personal protection may be obtained by using insect repellents.

FURTHER READING

Anderson, J.F. (1985) The control of horse flies and deer flies (Diptera: Tabanidae). *Myia*, **3**, 547–98.

Anthony, D.W. (1962) Tabanids as disease vectors. In *Biological Transmission of Disease Agents*. Symposium held under the auspices of the Entomological Society of America, Atlantic City, 1960, ed. K. Maramorosch, pp. 93–107. New York: Academic Press.

Duke, B.O.L. (1972) Behavioural aspects of the life cycle of *Loa*. In *Behavioural Aspects of Parasite Transmission*. ed. E.K. Canning and C.A. Wright, pp. 97–107. London: Academic Press.

Foil, L.D. (1989) Tabanids as vectors of disease agents. *Parasitology Today*, **5**, 88–95.

Krinsky, W.L. (1976) Animal disease agents transmitted by horse flies and deer flies (Diptera: Tabanidae). *Journal of Medical Entomology*, **13**, 225–75.

8

Tsetse-flies (Glossinidae)

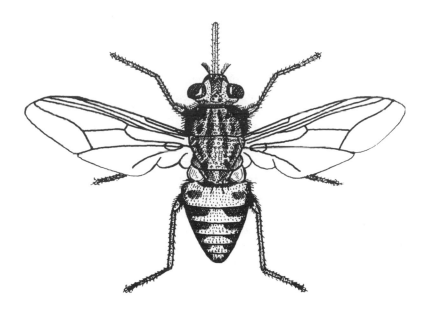

There are 23 species of tsetse-flies, six of which are divided into two or more subspecies, all of which belong to the genus *Glossina*. Apart from two localities in the Arabian peninsula, tsetse-flies are restricted to sub-Saharan Africa from between approximately latitudes 10° north and 20° south, but extending 30° south along the eastern coastal area. Some species such as *Glossina morsitans* are found across West Africa to Central and East Africa, whereas others are more restricted in their distribution. For example, *G. palpalis* occurs in only the West African subregion.

Tsetse-flies are vectors of both human and animal African trypanosomiasis, the disease in humans being referred to as sleeping sickness. The most important species are *G. palpalis*, *G. tachinoides*, *G. fuscipes*, *G. pallidipes* and *G. morsitans*.

8.1   EXTERNAL MORPHOLOGY

A general description of tsetse-flies, without special reference to any particular species, is as follows. Adults are yellowish or brown–black robust flies that are rather larger (6–14 mm) than house-flies. In some species the abdominal segments are uniformly coloured whereas in others there may be lighter-coloured transverse stripes and a median longitudinal one. Tsetse-flies are readily distinguished from all other biting flies and similar-sized non-biting flies by the combination of a rigid and forwardly projecting proboscis and a characteristic wing venation. Between veins 4 and 5 there is a closed cell which, with a little imagination, looks like an upside-down hatchet (i.e. axe, cleaver or chopper), and is consequently often termed the hatchet cell (Figs. 8.1*b*, 8.2*a*). This one character serves to identify a fly as a tsetse. Tsetse-flies also differ from most flies in that at rest the wings are placed over the abdomen like the closed blades of a pair of scissors (Fig. 8.1*a*).

A long pair of palps arise dorsally, very close to the proboscis, and lie

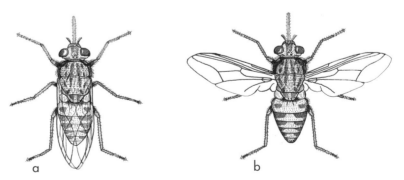

Figure 8.1 Tsetse-fly. (*a*) Wings folded over body like a pair of closed scissors; (*b*) wings spread out to display abdomen and wing venation.

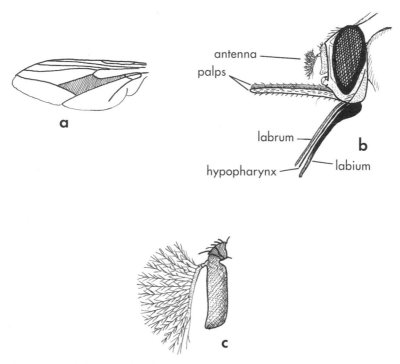

Figure 8.2 Adult tsetse-fly. (*a*) Wing showing shaded hatchet cell; (*b*) lateral view of head showing mouthparts, antenna and palps; (*c*) antenna showing plumose branching of hairs on arista.

alongside it. They are difficult to distinguish except when the tsetse-fly is feeding and the proboscis is swung downwards while the palps remain projecting forwards (Fig. 8.2*b*). The first two antennal segments are small and inconspicuous but the third is relatively large and cylindrical and is also somewhat banana-shaped. This has near its base the arista which has branched hairs, but only on the upper surface, giving it a feathery appearance (Fig. 8.2*c*).

The proboscis is relatively large and has a bulbous base. When a tsetse feeds saliva containing anticoagulants is pumped down into the wound formed by the fly.

The dorsal surface of the thorax of a tsetse-fly has a pattern of dark brown stripes or patches. There are eight abdominal segments and these may be uniformly dark brown or blackish, or have transverse stripes and a median one of a lighter brown or yellowish colour. Although, as far as disease transmission is concerned, it may not be important to distinguish between the sexes because both take blood-meals; this can nevertheless be done easily by examining the tip of the ventral surface of the abdomen. In the male tsetse-fly there is a prominent raised, almost circular, button-like structure called the hypopygium, which when unfolded reveals a pair of genital claspers. In the female fly there is no such button-like protuberance.

8.1.1   Alimentary canal of the adult fly (Fig. 8.3)

A knowledge of the morphology of the alimentary canal and associated salivary glands is essential for an understanding of the life cycle of trypanosomes within the tsetse.

The food channel, formed by the apposition of the labrum and labium, leads to the pharynx and then to the oesophagus which has a slender duct leading to an oesophageal diverticulum, commonly called the crop. Just behind the oesophagus there is an important bulbous structure termed the proventriculus. The distal end of the proventriculus marks the end of the fore-gut and beginning of the mid-gut, which in the tsetse-fly is very long and convoluted. A peritrophic membrane, which plays a part in the cyclical development of sleeping sickness trypanosomes (*Trypanosoma brucei gambiense, T. brucei rhodesiense*) in the tsetse-fly, is secreted by the epithelial cells in the anterior part of the proventriculus. The peritrophic membrane when first produced by the proventriculus is a very delicate, soft and almost fluid structure, but as it passes back into the gut it hardens to form a thin but rel-

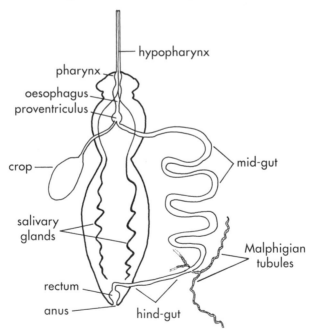

Figure 8.3 Diagrammatic representation showing the alimentary canal of a tsetse-fly and the thread-like salivary glands.

atively tough sleeve, something like a sausage skin. This tube-like peritrophic membrane lines the entire length of the mid-gut.

The junction of the four Malpighian tubules separates the mid-gut from
the hind-gut, which terminates in a small dilated rectum and opens to the
exterior through the anus.

The slender paired salivary glands originating in the head of the tsetse
are enormously long, very convoluted and stretch back to almost the end of
the abdomen. Anteriorly, the ducts from both glands unite in the head to
form the common salivary duct which passes down the length of the hypopharynx, which runs down the centre of the proboscis.

8.2   LIFE CYCLE

8.2.1   Feeding and reproduction

Both male and female tsetse-flies bite people, a large variety of domesticated and wild mammals, and also reptiles and birds. No species of tsetse
feeds exclusively on one type of host but most species show definite host
preferences. For example, in East Africa *Glossina swynnertoni* feeds mainly
on wild pigs and *G. morsitans* on wild and domesticated bovids as well as
on wild pigs, but in West Africa *G. morsitans* feeds mainly on warthogs. In
East Africa *G. pallidipes* feeds principally on wild bovids, whereas in West
Africa *G. palpalis* feeds predominantly on reptiles and humans, and *G. tachinoides* feeds on humans and bovids, but in southern Nigeria predominantly on domestic pigs. Tsetse-flies take blood-meals about every 2–3
days, although this interval may be reduced in dry, hot weather or prolonged for about 10 days in cool, humid conditions. Feeding is restricted to
daylight hours and vision plays an important part in host location, dark
moving objects being particularly attractive. On pale-skinned people, such
as Caucasians, tsetse-flies often bite through dark clothing such as socks,
trousers and shorts in preference to settling on the skin. During feeding
blood is sucked up the proboscis, passes to the crop and later to the mid-
gut where digestion proceeds.

The different types of flies so far described in this book lay eggs; in
marked contrast tsetse-flies do not, and instead they deposit larvae one at a
time. (Adults of *Sarcophaga* and *Wohlfahrtia* also deposit larvae not eggs
(Chapter 10).)

After a female tsetse-fly has been inseminated by a male and after it has
taken a blood-meal, a single egg in one of the two ovaries completes maturation. It then passes down the common oviduct into the uterus where it is
fertilized by the release of spermatozoa from the spermathecae. The egg
hatches within the uterus after about 3–4 days, and the empty eggshell
passes out through the genital orifice (vagina). The uterus is supplied with

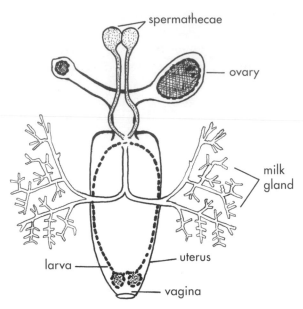

Figure 8.4 Diagrammatic representation of the female reproductive system of a tsetse-fly, with a full-grown larva in the uterus.

a conspicuous pair of branched secretory accessory glands which in tsetse-flies are called the milk glands (Fig. 8.4). Fatty, nutrient fluid from these glands flows through a small duct to enter the uterus at its anterior end. The larva is orientated within the uterus so that its mouth is near the opening of the common duct of the milk glands. The secretions from these glands provide the larva with all the food it needs for growth and development. The larva passes through three instars in the female. Regular blood-meals must be taken by the female for a continuous and adequate provision of nutrient fluid from the milk glands. If the fly is unable to feed, the larva may fail to complete its development and as a consequence be 'aborted'.

Larval development is completed after about 9 days, by which time the third and final instar larva is 5–7 mm long. It is creamy white in colour and composed of 12 visible segments, the last of which bears a pair of prominent dark protuberances called the polypneustic lobes (Fig. 8.5b), which are respiratory structures. A female containing a fully developed larva is easily recognized because the fly's abdomen is enlarged and stretched, i.e. the fly is obviously 'pregnant'. Furthermore, the black polypneustic lobes can be seen through her abdominal integument.

The mature third-instar larva wriggles out, posterior end first, from the genital orifice; thus birth can be termed a 'breech case'. Females always

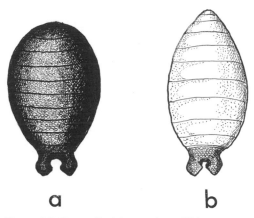

Figure 8.5 Tsetse-fly. (*a*) Puparium; (*b*) larva.

select shaded sites for larviposition. The larva is deposited on loose friable soil, sand or humus, frequently underneath bushes, trees, fallen logs, rocks, between buttress roots of trees, in sandy river beds, in animal burrows and even in rot-holes in trees which may be formed some distance above the ground (4–5 m). Immediately the larva is deposited it commences to bury itself under 2–5 cm of soil. After about 15 minutes the third-instar larval skin contracts and hardens to form a reddish-brown or dark brown, barrel-shaped puparium which is about 5–8 mm long and has distinct polypneustic lobes (Fig. 8.5*a*). Within this puparial case the larva pupates.

The duration of the puparial period is comparatively long, usually extending over 4–5 weeks, but at high temperatures (30 °C) may be completed within 3 weeks, and conversely at low temperatures (20 °C) prolonged to 7 weeks. After puparial development has been completed the fly emerges from the puparium, forces its way to the surface of the ground and flies away.

During the development of the larva within the female the tsetse feeds several times, about once every 2–3 days. The first larva is deposited about 16–20 days after the female has emerged from the puparium; thereafter, if food is plentiful a larva is deposited about every 9–12 days. In the laboratory female tsetse-flies have produced up to 20 offspring but the average is nearer 5–8. Breeding generally continues throughout the year but in very humid conditions reproduction may be diminished. Maximum population size is usually attained at the end of the rainy season. The population diminishes in the dry season when suitable areas of refuge for adult flies and suitable larviposition sites may become restricted and localized.

8.2.2   Adult behaviour

Knowledge of certain aspects of the behaviour of tsetse-flies is essential for an understanding of approaches to their control, and also the part vector species play in the transmission of sleeping sickness.

Blood-engorged tsetse-flies, and unfed hungry flies waiting to feed on suitable hosts, spend the nights and much of the daytime hours resting in dark and usually humid sites. In fact tsetse-flies spend about 23 hours a day resting on vegetation. During the day the favoured resting sites of most species are twigs, branches and trunks of trees and bushes. Flies are not found resting in sites in which temperatures rise above about 36 °C. At night tsetse-flies prefer to rest on the upper surfaces of leaves. Accurate knowledge of the actual resting sites may be required for control measures. For example, the height at which adults rest on trees determines the height at which the trees need to be sprayed with insecticides. Most species in fact rest below 4 m; in Nigeria 50% of *G. palpalis* and *G. morsitans* commonly rest between ground level and 30 cm.

Based on their morphology, ecology, karyotype and behaviour tsetses can be separated into the following three main groups:

Fusca group (forest flies)

The *fusca* group contains 15 species and subspecies of *Glossina*, all of which are large (10.5–15.5 mm). They are forest flies (except *G. longipennis*, which occurs in arid areas of East Africa) and most are restricted to the equatorial forests of West and West Central Africa. For example, *G. fusca* occurs mainly in relict forests of West and Central Africa, whereas *G. brevipalpis* is found in secondary forests of East Africa.

The *fusca* group rarely feeds on people and none of the species is a vector of sleeping sickness.

Morsitans group (savannah flies)

Seven species and subspecies are included within the *morsitans* group. They are medium-sized insects, 7.5–11 mm long, and typically inhabit the savannah regions of Africa, which may extend from the coast, or the edges of forests to dry semidesert regions. *Glossina morsitans* occupies the savannah regions of West, Central and East Africa, whereas *G. pallidipes* is found in savannahs of East Africa and parts of southern Africa, and *G. swynnertoni* is restricted to the savannahs of a very limited area of East Africa. *Glossina morsitans* and *G. pallidipes* occur in areas ranging from wooded savannah at the edges of forests to the dry thicket vegetation of arid zones, whereas *G. swynnertoni* is restricted mainly to relatively dry thicket country.

All three species above are vectors of sleeping sickness, but the most important is *G. morsitans*.

Palpalis group (riverine and forest flies)

Nine tsetse species and subspecies are included in the *palpalis* group, the smallest being about 6.5 mm in length and the largest 11 mm. They are essentially flies inhabiting wetter types of vegetation, such as forests, luxuriant scrub and vegetation growing along rivers and shores of lakes. *Glossina palpalis* inhabits riverine vegetation bordering rivers and lakes, mangrove swamps and forested areas, and occurs throughout most of West Africa, down the western part of the continent to Angola. *Glossina fuscipes*, which is closely related to *G. palpalis*, occurs mainly in Central Africa but extends its range to the western areas of East Africa. *Glossina tachinoides* is a riverine species found near streams and rivers in wet humid coastal areas, through wooded savannah regions to the riverine vegetation of very dry savannah areas. It is found mainly in West and Central Africa but also occurs in parts of Ethiopia and Sudan. All these species are vectors of sleeping sickness.

8.3   MEDICAL IMPORTANCE

All species of *Glossina* are potential vectors of African trypanosomiasis to humans. In practice, however, relatively few species of tsetse-flies are natural vectors because many species rarely, if ever, feed on people. It is the behaviour of the adult tsetse and the degree of fly–human contact, and in the case of Rhodesian sleeping sickness also the degree of vector contact with the reservoir hosts of the trypanosomes, that establishes whether a tsetse-fly is a vector.

Sleeping sickness has a patchy distribution over about 10^6 km^2 of land in Africa; some 20000–25000 new cases are detected annually, but many more cases remain undetected. There are two subspecies of trypanosomes (previously often regarded as distinct species) causing sleeping sickness in humans, namely *Trypanosoma brucei gambiense* and *T. brucei rhodesiense*. These parasites are morphologically indistinguishable but produce different clinical symptoms and have different epidemiologies. The most important vectors of sleeping sickness are *G. palpalis*, *G. fuscipes*, *G. tachinoides*, *G. morsitans* and *G. pallidipes*.

The cycle of development of *Trypanosoma brucei gambiense* and *T. brucei rhodesiense* in the tsetse-fly is the same and is as follows. Trypanosomes in the blood are sucked up by male or female tsetse-flies during blood-feeding from an infected person (or, particularly in the case of *T. brucei rhodesiense*, often from a non-human reservoir) and pass first through the oesophagus to the crop, and then, after feeding has ceased, into the peritrophic tube lining the mid-gut. Blood is digested within the mid-gut but the *gambiense* and *rhodesiense* trypanosomes are not destroyed. About 9–11 days after feeding the trypanosomes penetrate the middle part of the peritrophic membrane and pass across into the space between the membrane and gut, called the ectoperitrophic space. Here they multiply and after about 3–9 days the parasites migrate to infect the proventriculus by penetrating the anterior softer part of the peritrophic membrane. From here they pass down the food channel in the proboscis and pass up the hypopharynx or salivary duct to invade the salivary glands, where they develop into epimastigotes and multiply enormously. About 15–35 days after an infective blood-meal the tsetse-fly becomes infective and metacyclic trypomastigotes are injected into a vertebrate host when the fly feeds. (It remains unclear whether there are 'short-cuts' to this cycle of development – for example the possibility that parasites penetrate the gut and pass across into the haemocoel from where they can migrate directly to the salivary glands.)

When determining infection rates in tsetse-flies by dissection, any trypanosomes found in the gut or proboscis are ignored as only those in the salivary glands can be ascribed with any degree of certainty to *T. brucei gambiense* or *T. brucei rhodesiense*, but there are complications: *T. brucei brucei*, which does not cause sleeping sickness in humans but causes an animal trypanosomiasis commonly called nagana, undergoes a similar cyclical development in the fly. Consequently, the presence of trypomastigotes (metacyclic forms) in the salivary glands does not necessarily indicate the presence of trypanosomes infective to people.

Salivary gland infection rates in tsetse-flies, for some incompletely known reasons, are low, rarely exceeding 0.1%, even in endemic areas.

8.3.1   Gambian sleeping sickness

Gambian sleeping sickness is a form of the disease that occurs from West Africa through Central Africa to parts of Sudan and southwards to Angola and Zaire. *Glossina palpalis* and *G. tachinoides* are the most important vectors in West Africa, and *G. fuscipes* is the vector in Central and East Africa, of *T. brucei gambiense*, the causative agent of Gambian sleeping sickness. This disease is relatively chronic, with death often not occurring until after many years. Until recently it was considered that there were no natural reservoirs of the disease other than humans, but studies in West Africa indicate that domestic pigs and some wild mammals may harbour *T. brucei gambiense*.

The vectors of Gambian sleeping sickness are especially common at watering places, fords across rivers and along lake shores, etc., in fact in places where people frequently visit to collect water or do their washing. As a consequence there may be limited and localized foci of transmission.

8.3.2   Rhodesian sleeping sickness

The causative agent of Rhodesian sleeping sickness, *T. brucei rhodesiense*, causes a more virulent disease than *T. brucei gambiense* but it is not so widespread, being more or less restricted to Tanzania, Malawi, Zambia, Zimbabwe, Mozambique and to the northern areas of Lake Victoria in Kenya and Uganda. The most important vectors are *G. morsitans* and *G. pallidipes*, species which feed on a variety of game animals and domestic livestock, especially bovids, in preference to people. These flies often occur in savannah areas thinly populated by humans. Wild animals, especially a number of bovid species, are important reservoirs of *T. brucei rhodesiense*. Around Lake Victoria, *G. fuscipes* is the main vector and cattle are an important reservoir of disease. The disease is a zoonosis.

8.4   CONTROL

Because the immature stages are so well protected, the larva being retained by the female for almost all of its life, and the puparium being buried in the soil, control of tsetse-flies is aimed at the adults.

Many control methods have been directed at tsetse-flies to combat human and animal trypanosomiasis. At one time there were campaigns in Zimbabwe and some East African countries to kill in selected areas all game animals that might provide food for tsetses, or be reservoirs of trypanosomiasis. The wide-scale and often indiscriminate slaughter of animals is no longer acceptable in a world that is increasingly sensitive to the preservation of wildlife.

The distribution and abundance of tsetse-flies is largely determined by types of vegetation and microclimate. Their dependence on such requirements formed the basis of control through clearing away vegetation, especially near rivers and lake shores, either completely or partially. The destruction of such vegetation in the past has achieved considerable success in controlling tsetse-flies, but in many countries this method, at least on a large scale, is no longer ecologically or economically acceptable. However, clearing vegetation for human settlements or felling trees for firewood may also have the same effect in reducing tsetse-fly populations.

8.4.1   Insecticidal control

There are no problems of insecticide resistance and tsetse-fly control relies mainly on insecticides. These can be applied as persistent chemicals that will remain effective in killing flies for at least 2–3 months, or as non-residual aerosols that kill when the droplets land on resting or flying tsetse-flies. In the former and cheaper approach, wettable powders or emulsions of insecticides such as DDT, dieldrin, endosulfan, permethrin, deltamethrin and cypermethrin can be sprayed onto vegetation harbouring adults. The success of such methods depends on a detailed knowledge of the behaviour and resting sites of the vectors. For example, it is often possible to restrict spraying of tree trunks up to a height of 1.5 m during the dry season, extending this to 3.5 m in the wet season. Insecticides are often sprayed on vegetation at concentrations of 1.5–5% active ingredient, either from knapsack sprayers or from other ground-based machines. The frequency of application depends much on local conditions, such as rainfall, other climatic factors and the growth and spread of new vegetation, which can provide new resting sites free of insecticides. Spraying is often done in the dry season so that deposits are not washed from vegetation by rainfall.

Alternatively, ultra-low-volume (ULV) aerial spraying of residual insecticides, using fixed-wing aircraft or helicopters, can be done. Although such aerial spraying results in greater coverage of vegetation with insecticides than can be achieved with ground application techniques, it usually kills a greater number of non-target insects and animals, many of which may be directly or indirectly beneficial to the human population.

Non-residual aerosols of insecticides such as endosulfan, or pyrethroids such as deltamethrin and lambdacyhalothrin (Icon), are usually made from aircraft. Repeated applications, such as about five sprayings each separated by 10 day intervals, are necessary because the tsetse-fly population is in the soil as puparia for 4–5 weeks or longer, and is unaffected by such spraying.

There is no doubt that indiscriminate or intensive and repeated spraying of residual organochlorine insecticides can have disastrous effects on the local fauna, but it should be appreciated that the application rates of insecticides for tsetse-fly control are often very low. Furthermore, after flies have been controlled in an area and spraying stopped, many, if not all, of the wildlife may revert to their original population numbers. Intensive bush clearance for tsetse-fly control can also greatly diminish the numbers and variety of the local fauna, but in this instance the change may be permanent, especially if people occupy and farm areas that were originally scrub or forest.

8.4.2   Targets and traps

Increasing use is made of more environmentally friendly and often cheaper techniques such as employing targets and traps. Targets, sometimes called screens, are coloured blue or black, are about 1 m² and are mounted on poles or hung from trees. They are very attractive to tsetse-flies and when they are impregnated with pyrethroids (e.g. deltamethrin, lambdacyhalothrin (Icon) or cyfluthrin) they kill adult flies settling on their surfaces. Re-impregnation is needed every 3–4 months.

Biconical or pyramidal traps made partially from dark blue or black material can be remarkably effective in attracting and trapping flies. Because some flies fail to enter traps, but settle on their outside surfaces, traps are often impregnated with pyrethroid insecticides to kill the potential escapees. Traps should be retreated with insecticides about every 6–10 months. The location of both screens and traps is critical for efficient control; generally open and sunny sites are best as they offer good visibility for the flies.

Recently there has been a move away from government-organized tsetse control campaigns to operations, using screens and traps, implemented by local people. Success of such community participation for the control of sleeping sickness has been mixed; it has usually proved difficult to sustain initial enthusiasm and success once outside inputs are removed.

8.4.3   Genetic control

Although genetic control, mainly using sterile male release (see pp. 24–5) methods, has sometimes been successful against tsetse-fly vectors of cattle trypanosomiasis, even leading to the eradication of flies from Zanzibar, the approach is not feasible for sustained control of human sleeping sickness.

FURTHER READING

Buxton, P.A. (1955) The natural history of tsetse-flies. An account of the biology of the genus *Glossina* (Diptera). *London School of Hygiene and Tropical Medicine Memoir*, **10**. London: H.K. Lewis.

Challier, A. (1982) The ecology of tsetse (*Glossina* spp.) (Diptera: Glossinidae): a review (1970–1981). *Insect Science and its Application*, **3**, 97–143.

Dransfield, R.D., Williams, B.G. and Brightwell, R. (1991) Control of tsetse flies and trypanosomiasis: myth or reality? *Parasitology Today*, **7**, 287–91.

Ford, J. (1971) *The Role of Trypanosomiases in African Ecology: A Study of the Tsetse Fly Problem*. Oxford: Clarendon Press.

Jordan, A.M. (1985) Tsetse eradication plans for southern Africa. *Parasitology Today*, **1**, 121–3.

Jordan, A.M. (1986) *Trypanosomiasis Control and African Rural Development*. London and New York: Longman.

Jordan, A.M. (1989) Man and changing patterns of the African trypanosomiases. In *Demography and Vector-Borne Diseases*, ed. M.W. Service, pp. 47–58. Boca Raton, Florida: CRC Press.

Laveissière, C. (1988) XV. Biology and control of *Glossina* species, vectors of human African trypanosomiasis. WHO/VBC/88.958. Geneva: World Health Organization mimeographed document.

Laveissière, C., Vale, G.A. and Gouteaux, J.P. (1991) Bait methods for tsetse control. In *Appropriate Technology in Vector Control*, ed. C.F. Curtis, pp. 47–74. Boca Raton, Florida: CRC Press.

Leak, S.G.A. (1999) *Tsetse Biology and Ecology: Their Role in the Epidemiology and Control of Trypanosomiasis*. Wallingford: CABI Publishing.

Mulligan, H.W. (ed.) (1970) *The African Trypanosomiases*. London: George Allen and Unwin.

Nash, T.A.M. (1969) *Africa's Bane: The Tsetse Fly*. London: Collins.

World Health Organization (1997) *Vector Control: Methods for Use by Individuals and Communities*, prepared by J.A. Rozendaal. Geneva: World Health Organization.

World Health Organization (1998) Control and surveillance of African trypanosomiasis. *World Health Organization Technical Report Series*, **881**, 1–113.

9

House-flies and stable-flies (Muscidae) and latrine flies (Fanniidae)

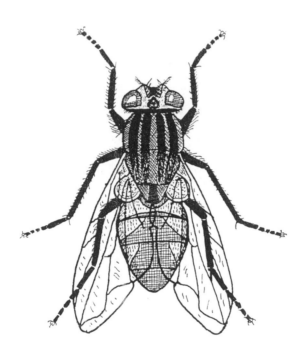

9.1   MUSCIDAE: HOUSE-FLIES AND STABLE-FLIES

There are many genera and about 3900 species of flies belonging to the family Muscidae. The most important from the medical aspect are the common house-fly, other house-flies and the stable-fly, all of which have a more or less world-wide distribution. The house-flies can be vectors of helminths, faecal bacteria, protozoans and viruses, resulting in the spread of such enteric diseases as the dysenteries and typhoids, whereas the stable-fly can cause a biting nuisance.

9.2   THE COMMON HOUSE-FLY (*MUSCA DOMESTICA*)

9.2.1   External morphology

There are about 66 species of flies in the genus *Musca*. The most common is *M. domestica*, the house-fly, which occurs in all parts of the world but is least common in Africa, where it is largely replaced by two subspecies (*M. domestica curviforceps* and *M. domestica calleva*). Other important related species include *Musca sorbens*, which can be a great nuisance in Africa, Asia and the Pacific; the notorious troublesome bush-fly of Australia, namely *M. vetustissima*; and *M. autumnalis*, which is a pest in both the Old and New Worlds. The appearance and biology of these *Musca* species are very similar. The morphology and biology of the house-fly (*M. domestica*) are described here.

House-flies are medium sized non-metallic flies about 6–9 mm in length, varying in colour from light to dark grey. They have four broadish dark longitudinal stripes on the dorsal surface of the thorax (Fig. 9.1a). The antennae consist of three segments, the distal and largest of which is cylindrical and has a prominent hair, called an arista, which has hairs on both sides. The antennae are concealed in depressions on the front of the face of the fly and are not easily seen. The mouthparts (proboscis) of the house-fly are complicated and specially adapted for sucking up fluid or semifluid foods. When not in use they are partially withdrawn into the head capsule (Fig. 9.2a), but are extended vertically downwards in a telescopic fashion when the fly feeds (Fig. 9.2b). The proboscis ends in a pair of oval-shaped fleshy labella, having very fine channels called pseudotracheae through which fluids can be sucked up. House-flies feed on a great variety of substances, such as sugar, milk, almost all food of humans, rotting vegetables and carcasses, excreta and vomit – in fact almost any organic material. The method of feeding differs according to the physical state of the food. For example, for thin fluids, such as milk and beer, the labella are closely adpressed on the food which is then sucked up through the small openings in the pseudotracheae. When feeding on semisolids such as excreta, sputum and nasal discharges, the labella are completely everted and food is sucked up directly into a food channel formed by the

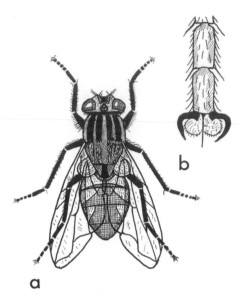

Figure 9.1 *Musca domestica*. (*a*) Adult fly; (*b*) terminal tarsal segments showing paired claws, paired large pulvilli and single bristle-like empodium.

apposition of the slender labrum and blade-like hypopharnyx. If flies feed on more solid materials such as sugar lumps, dried blood, cheese and cooked meats, the labella are everted and minute prestomal teeth surrounding the food channel are exposed and scrape away at the solid food. The fly then moistens small particles with either saliva or the regurgitated contents of its crop, after which the food is sucked up. This latter type of feeding is clearly a method which is conducive to the spread of a variety of pathogens.

The wings of the house-fly have vein 4 bending up sharply to join the costa close to vein 3 (Fig. 9.3). This is an important taxonomic character which can help distinguish *Musca* species from other rather similar flies. Each of three pairs of legs ends in a pair of claws and a pair of fleshy pad-like structures called the pulvilli, which are supplied with glandular hairs (Fig. 9.1*b*). These sticky hairs enable the fly to adhere to very smooth surfaces, such as windows, and are also responsible for the fly picking up dirt and pathogens when it visits excreta, septic wounds, rubbish dumps, etc. There are four visible greyish abdominal segments which are usually partially obscured from view by the wings.

9.2.2   Life cycle

Female *Musca domestica* are attracted to a variety of decomposing materials for egg laying, such as horse manure, poultry dung, urine-contaminated

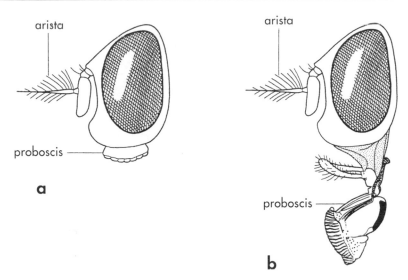

arista

arista

proboscis

a

proboscis

b

Figure 9.2 Lateral view of the head of *Musca domestica*. (*a*) Proboscis retracted; (*b*) proboscis extended for feeding.

bedding, foodstuff, carcasses, decomposing organic materials found in rubbish dumps, household garbage and waste foods from kitchens and hotels. During egg laying, 75–120 eggs are deposited together, or in separate batches, either in cracks and crevices or scattered over the surface. A fly may deposit five or six such egg batches in her lifetime. The eggs are creamy-white, 1–1.2 mm long, and distinctly concave dorsally giving them a banana-shaped appearance (Fig. 9.4*a*). They can hatch after only 6–12 hours, but this period is extended in cool weather. Hatching is accomplished by the strip of eggshell between parallel ridges on the dorsal concave surface lifting up, and partially detaching itself from the rest of the egg. Eggs cannot withstand desiccation and die if they dry out. Neither can they tolerate extremes of temperatures, most dying after exposure to temperatures below 15 °C or above 40 °C.

The creamy-white larvae which hatch from the eggs have a small head followed by an 11-segmented cylindrical and maggot-shaped body (Fig. 9.4*b*). At the pointed head end there is a pair of small curved mouthhooks, which are seen as a blackish structure situated beneath the integument of the head, and which are continued as the cephalopharyngeal skeleton in the anterior thoracic segments. At the posterior end of the body there is a pair of conspicuous spiracles shaped like a letter D. Each spiracle has a complete and thick outer wall called the peritreme which encloses three very sinuous spiracular slits (Fig. 9.4*c*).

Larvae feed on liquid food resulting from decomposing and decaying organic material. There are three larval instars. Mature larvae measure

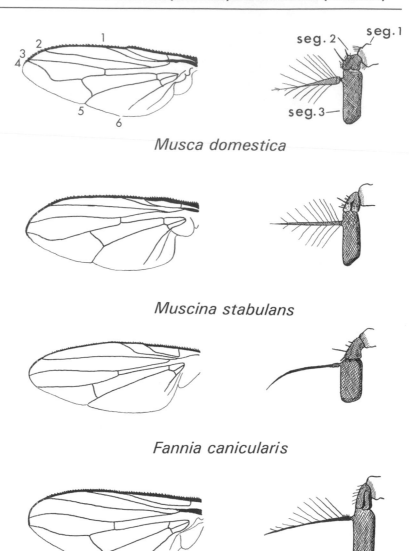

Musca domestica

Muscina stabulans

Fannia canicularis

Stomoxys calcitrans

Figure 9.3 Wing venation and antennae characteristic of the genera *Musca*, *Muscina*, *Fannia* and *Stomoxys*. Note the endings of veins 3 and 4.

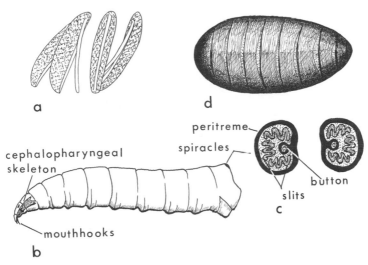

Figure 9.4 *Musca domestica*. (*a*) Two hatched and two unhatched eggs; (*b*) final instar larva; (*c*) posterior D-shaped larval spiracles; (*d*) puparium.

about 8–14 mm, the final size depending on environmental conditions, especially the amount of available food. The speed of larval growth and development depends on the abundance of food supply and temperature. Development may be completed within only 3–5 days, but under less favourable conditions 7–10 days are needed, and in cool weather development may extend to about 24 days.

If larval habitats dry out the larvae are killed, but if they become too wet they drown.

Prior to pupation the third-instar larvae may migrate away from their larval habitats to drier ground. Sometimes, however, the periphery of breeding places, such as rubbish dumps, may be sufficiently dry for larvae to pupate there; pupation may also occur in the dry soil underneath larval habitats. Pupation begins with the larval skin contracting, hardening and turning dark brown, during which a barrel-shaped structure measuring about 6 mm, called a puparium, is formed. Close examination shows this is segmented (Fig. 9.4*d*). The puparium is often referred to as the pupa but technically this is incorrect because the actual pupa is formed within the protective shell of the puparial case. (A puparium is also formed by tsetse-flies (Chapter 8) and by myiasis-causing flies (Chapter 10).) The puparial stage lasts about 3–5 days in warm weather but may be prolonged to 7–14 days during cooler periods.

Developmental time from egg to adult (Fig. 9.5) is about 49 days at 16 °C, 25 days at 20 °C, 16 days at 25 °C, 10–12 days at 30 °C and 8–10 days at 35 °C.

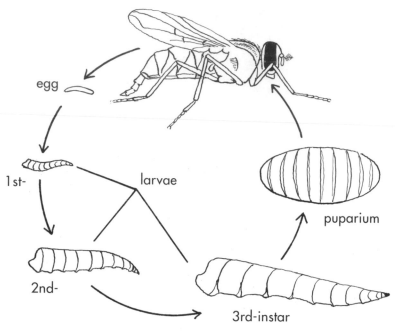

Figure 9.5 Life cycle of *Musca domestica*, which is typical of other muscid and also calliphorid (see Chapter 10) flies.

Very occasionally the period can be less than 7 days. In temperate areas a varying, but usually small, proportion of house-flies survive throughout the winter as puparia, but more frequently they overwinter as hibernating adults. Immature development ceases at temperatures below 12 °C and a temperature of about 45 °C is lethal to the eggs, larvae and puparia.

The adult fly escapes from its puparial case by pushing off its anterior end and crawling out. After hardening of the cuticle and inflation of the wings, the adult flies away.

Adults of *Musca domestica* generally avoid direct sunlight, preferring to seek shelter in buildings inhabited by people or their animals. House-flies and related flies, and the calliphorid flies discussed in Chapter 10, are often called domestic or synanthropic flies, because of their close association with humans and their homes. House-flies defaecate at random, and frequently regurgitate their food, resulting in unsightly 'fly spots'. Adults tend to stay within 500 m of their breeding sites, but can fly 1–5 km, and sometimes even greater distances.

Musca sorbens is widely distributed in Africa and parts of Asia, where it is a greater pest than *M. domestica* because adults more commonly settle on suppurating wounds, around the eyes, sweaty skin and other body secretions than does the common house-fly.

9.2.3   Medical importance

House-flies can transmit a large number of diseases to humans owing to their habits of visiting, almost indiscriminately, faeces and other unhygienic matter and people's food. Pathogens can be transmitted by three possible routes:

1. By flies' contaminated feet, body hairs and mouthparts. Most pathogens, though, remain viable on the fly for less than 24 hours, and there are usually insufficient numbers to cause a direct infection, except possibly with *Shigella*. However, if pathogens are first transferred to food they may then multiply sufficiently to reach the level of an infective dose.
2. By flies vomiting on food during feeding, which they do frequently.
3. By defaecation, which often occurs on food. This is probably the most important method of transmission.

Over 100 different pathogens have been recorded from house-flies, at least 65 of which are known to be transmitted. With the exception of *Thelazia* species transmission is mechanical, that is the fly acts just as a physical carrier.

House-flies can transmit viruses of polio, trachoma, Coxsackie virus and infectious hepatitis, as well as rickettsiae such as Q fever (*Coxiella burneti*), and numerous bacterial diseases, but mainly enteric ones, such as bacillary dysentery (*Shigella*), cholera, enterotoxic *Escherichia coli* (ETEC), *Campylobacter*, the typhoids and paratyphoids (*Salmonella*), as well as a variety of streptococci and staphylococci. They may also be vectors of protozoan parasites such as those causing amoebic dysenteries (*Entamoeba*, *Giardia*). In addition, house-flies can carry eggs and cysts of a variety of helminths for example *Taenia*, *Hymenolepis*, *Dipylidium*, *Diphyllobothrium*, *Necator*, *Ancylostoma*, *Thelazia*, *Enterobius*, *Trichurus* and *Ascaris*. Furthermore, they can be carriers in the tropical Americas of the eggs of *Dermatobia hominis*, a myiasis-producing fly (Chapter 10).

Larvae of house-flies have occasionally been recorded in cases of urogenital and traumatic myiasis, and more rarely in aural and nasopharyngeal myiasis. If food infected with fly maggots is eaten, then they may be passed more or less intact in the excreta, often causing considerable alarm and surprise. There is, however, no true intestinal myiasis in humans (Chapter 10).

Because of their dirty habits house-flies are potential vectors of many pathogens to humans, but it is difficult to assess their relative importance in the transmission of most diseases. Much information of their real role in the spread of disease has been circumstantial; for example, seasonal increase of fly abundance is often closely correlated with outbreaks of diarrhoeal diseases. The classic demonstration of the association between flies and disease was in two Texan towns in 1946 and 1947. One town was sprayed with DDT to destroy the house-flies and this was accompanied by a reduction in acute diarrhoeal diseases and deaths in children due to *Shigella*, although *Salmonella* infections remained the same. The unsprayed town did not have such a reduction in *Shigella* infections. More recently trials involving house-fly control seem to support the belief that under certain circumstances they may be important disease vectors. For example, when in 1995 and 1996 breeding sites in Pakistan villages were sprayed with insecticides fly populations decreased by 97% and the incidence of childhood diarrhoea decreased by 23%. Similarly in The Gambia ultra-low-volume (ULV) spraying with deltamethrin in 1997 and 1998 reduced fly populations by 87%, the prevalence of trachoma was 4 times lower in villages that had been sprayed than in those that were not, and diarrhoea in children was reduced by 25%. In 1988 control of house-flies by attractant traps in an Israeli army camp resulted in a reduction in *Shigella* infections and also apparently ETEC infections. There is now considerable evidence that house-flies contribute to ill-health.

9.2.4   Control
Control methods can be divided conveniently into three categories.

1. Physical and mechanical
2. Environmental sanitation
3. Insecticidal

Physical and mechanical control
Flies and other obnoxious insects can sometimes be prevented from entering buildings by screening windows, openings, air vents, etc. Screens should be made of non-corrosive material, such as plastic, copper or aluminium gauze, and have a mesh size of 3–4 strands per centimetre. This will exclude relatively large insects such as house-flies from buildings without unduly decreasing air circulation or light. Screening can be costly, but may be worthwhile in hospitals and restaurants. Screening should be periodically inspected and any tears mended. Screening can reduce fly nuisance but does not solve the problem, for flies will continue to breed locally and to enter unscreened houses.

The establishment in doorways of an air current, such as the air barriers found in the entrances of some shops, and fans mounted over doorways, may help to reduce the number of flies entering premises. The well-known practice of placing in doorways curtains made of many vertical, often coloured, strips of plastic or beading also helps to keep out flies. A common method to kill flies that enter restaurants and food stores, is to mount an ultraviolet light-trap on a wall. Flies attracted to the trap are electrocuted on entering it by contact with an electric grid.

Environmental sanitation

Environmental sanitation aims at dramatically curtailing house-fly populations by reducing their breeding places. For example, domestic refuse and garbage should be placed either in strong plastic bags and the openings tightly closed, or in dustbins with tight-fitting lids. When possible there should also be regular refuse collections, preferably twice a week in warm countries, to prevent any eggs laid among the garbage developing into adults. If household refuse cannot be collected it should be burnt or buried. Unhygienic rubbish dumps, so commonly found in towns and villages, provide ideal breeding places for house-flies, and should be removed. Refuse should be placed in pits which are covered, daily if possible, with a 15 cm layer of earth; when the pits are more or less full they must be finally covered with 60 cm of compacted earth. This depth of final fill is required to prevent rodents being attracted to buried decomposing organic material and burrowing into the rubbish pits.

It is also important to organize efficient disposal of household and industrial sewage and sanitary wastes. There should also be efficient latrines which prevent the breeding of house-flies and allied flies.

Insecticidal control

In many parts of the world house-flies have become resistant to many of the organochlorine insecticides, and these are now rarely used in control. In many areas resistance has also developed to the organophosphates so that other insecticides such as the carbamates or the pyrethroids have to be used, but unfortunately some house-fly populations have also become resistant to these chemicals. Research is being carried out on the use of *Bacillus thuringiensis* subsp. *israelensis* as a larvicide, adulticide and in attractant baits.

Larvicides Insecticides can be directed against the larvae, such as by spraying the insides of dustbins and refuse and garbage heaps, manure piles and other breeding sites with solutions or emulsions of organophosphate insecticides such as diazinon, bromophos, fenchlorphos (fenchlorvos, Ronnel), fenitrothion (Sumithion) or fenthion (Baytex), carbamates such as propoxur (Baygon) or carbaryl (Sevin), or insect growth regulators (IGRs) such as diflubenzuron (Dimilin), cryomazine (Neporex) or pyriproxyfen. A problem with any larvicide is that usually large volumes (0.5–5 litres/m^2) need to be applied to penetrate the upper 10–15 cm of the breeding site and reach the larvae.

The outside of dustbins and any adjacent walls should also be sprayed to deter gravid (egg-laden) flies from ovipositing in dustbins and other nearby breeding places.

Spraying against adults Many different types of insecticides and procedures have been used to reduce adult flies. For indoor use commercially
available aerosol dispensers, hand sprayers or larger but portable sprayers
can be used to spray knock-down insecticides to kill adult flies. Suitable
insecticides include 0.5% dichlorvos (DDVP), 2–4% malathion, 1% naled
(Dibrom), 2% pirimiphos methyl (Actellic), 0.1–0.5%, fenchlorphos
(Ronnel), 0.1–0.4% pyrethrins plus a synergist such as 0.5–2.5% piperonyl
butoxide, or 0.1% pyrethroids (e.g. permethrin, cypermethrin, deltamethrin) which do not need synergists and are about 20 times more toxic to
house-flies than natural pyrethrins. Aerosol applications and space-spraying have virtually no residual effects and consequently treatments have to
be repeated to achieve control and this can be costly. Furthermore this
approach does little to alleviate the source of the fly nuisance. Care needs to
be taken not to contaminate food with insecticides.

Outdoor application of insecticidal aerosols or mists from special spraying machines, or aerial ULV applications can give effective control in
certain situations, such as in and around dairies, farms, markets and recreational areas. Various organophosphate insecticides and pyrethroids, such
as those used for indoor space-spraying, are suitable. Generally only temporary relief from flies is achieved, as outdoor spraying does not usually
kill indoor resting flies or prevent reinvasion of flies from outside the
treated area or from breeding places.

Flies may also be controlled by spraying indoor walls, ceilings, doors,
etc. with residual insecticides such as 1–2 g/m^2 of malathion, fenchlorphos
(Ronnel), fenitrothion (Sumithion) or pirimiphos methyl (Actellic); or
0.4–1.0 g/m^2 diazinon or naled (Dibrom); or 0.1–0.2 g/m^2 bendiocarb
(Ficam) or propoxur (Baygon); or 0.025–0.1 g/m^2 cypermethrin or permethrin; or 0.01–0.15 g/m^2 deltamethrin. These residual insecticides
should remain effective for 1–2 months, but much depends on local
circumstances and whether the walls are washed. The outside walls of
houses and cattle sheds may also be sprayed with residual insecticides, but
their duration of effectiveness will depend on several factors, such as
whether the surface deposit is rubbed off, or washed off by rain.

Insecticidal cords Insecticidal cords comprise cord or rope strips soaked in
insecticides such as diazinon, fenthion (Baytex), propoxur (Baygon),
cypermethrin or permethrin and dyed, preferably red, to alert people that
they are impregnated with insecticides. Flies are killed when they rest on
these cords hung up in dairies and houses.

Dichlorvos (DDVP)-impregnated resin strips which disseminate insecticidal vapours are commonly used to control flies in houses, restaurants,
hotel kitchens, dairies, etc. They can remain effective for 2–3 months, but
this depends on temperature and degree of ventilation in the rooms in
which they are used.

Commercially available sticky tapes ('fly-papers') incorporating sugar can be relatively effective, although unsightly, in attracting and catching flies.

Toxic baits Attractant fly baits have also been used to control house-flies. Sometimes simple traps are baited with an attractant such as a bowl of fermenting yeast and flies which enter the trap are prevented from escaping; in Israel such traps have caught several kilograms of flies over a few days.

More usually baits are combined with insecticides, such as sugar mixed with sand, bran or some other inert carrier and treated with 0.1–2% insecticides, which provides an attractive but lethal solid bait. Liquid baits commonly comprise 10% sugar dissolved in water plus 0.1–2% insecticides. The most commonly used insecticides in both dry and liquid baits are malathion, diazinon, dichlorvos (DDVP), naled (Dibrom), fenchlorphos (Ronnel), trichlorphos (Dipterex), propoxur (Baygon) and bendiocarb (Ficam). More simply 1–2% formaldehyde can be mixed with liquid baits. Sometimes the house-fly sex pheromone called muscalure is added to baits to make them more attractive. In some countries commercially prepared dry or liquid baits are available. Dry baits can be scattered or placed in wooden, plastic or metal trays; they need to be replaced about every 2 days. Liquid baits can be placed in a glass bottle which is inverted over a saucer-like receptacle so that as the bait evaporates more flows in from the reservoir, as in automatic feeders for poultry. Occasionally paint-on baits comprising molasses, syrup or sugar and a thickening agent, and the organophosphate insecticide trichlorphon (trichlorfon), are applied directly to structural surfaces (e.g. walls) or to strips suspended from ceilings or fastened to walls. Attractant baits can sometimes quickly reduce fly populations.

9.3   THE GREATER HOUSE-FLY (*MUSCINA STABULANS*)

9.3.1   External morphology
Muscina stabulans has a world-wide distribution and is commonly referred to as the greater house-fly. Adults are about 7–10 mm long, slightly larger than house-flies. They can be distinguished from both *Musca* and *Fannia* because vein 4 of the wing curves slightly, but distinctly, upwards towards vein 3 (Fig. 9.3), and from *Fannia*, but not *Musca*, by having hairs on both the upper and lower sides of the arista (Fig. 9.3).

9.3.2   Life cycle
Females scatter about 150–200 eggs more or less indiscriminately over the surface of decaying matter such as rotting fruits, vegetables and fungi, on cooked and raw meats especially if putrified, on carcasses and also on human and animal excreta. The eggs hatch after 1–2 days and the resultant larvae resemble the maggot-shaped larvae of the house-fly, but can be

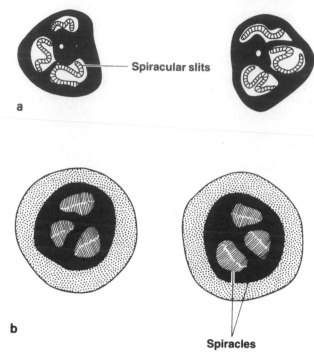

Spiracular slits

a

b

Spiracles

Figure 9.6 Posterior larval spiracles. (a) *Stomoxys calcitrans*; (b) *Muscina stabulans*.

distinguished by the structure of the posterior spiracles. In *M. stabulans* the spiracles are almost circular, not D-shaped as in *M. domestica*. Furthermore, the peritreme, which is very wide, encircles three crescent-shaped spiracular slits (Fig. 9.6b). Young larvae are omnivorous scavengers, as are larvae of *Musca* and *Fannia*, but towards the end of the larval period they become predacious, feeding on any other fly larvae in the breeding places. The brown puparium is similar in shape to that of *M. domestica* and its duration is about 1–2 weeks. Like *Fannia* species the life cycle is about 4–6 weeks, although in warm weather it may be reduced to 20–25 days.

Adults enter buildings and feed and behave much the same as do adults of *M. domestica*.

9.3.3   Medical importance

The exact role of *M. stabulans* as a disease vector remains undetermined, but many of the pathogens transmitted by the house-fly are probably also spread by *Muscina*. Larvae, like those of *Fannia* and *Musca*, have occasionally been recorded in cases of accidental intestinal myiasis.

9.3.4   Control

The control measures are very similar to those applied to the house-fly.

9.4   THE STABLE-FLY (*STOMOXYS CALCITRANS*)

9.4.1   External morphology

Stable-flies have a world-wide distribution. They are sometimes known as biting house-flies or, in the USA, as dog-flies (because they commonly bite dogs). The most common species is *Stomoxys calcitrans*. Adults have four black longitudinal stripes on a dark grey thorax and are about the size (5–6 mm) of house-flies, which they superficially resemble. They are, however, easily separated from *Musca*, *Fannia* and *Muscina* by a conspicuous forward-projecting, rigid proboscis (Fig. 9.7). The wing venation resembles that of *Muscina*. The arista of the third antennal segment differs from *Musca*, *Fannia* and *Muscina* in having hairs arising from only the upper side (Fig. 9.3).

In Africa adults might at first glance be confused with tsetse-flies, which also have a forward-projecting proboscis, but *Stomoxys calcitrans* is a smaller fly. Also, when at rest its wings are not placed completely over the body in a closed scissor-like fashion as in tsetse-flies, but are kept apart as in house-flies. Furthermore, there is no enclosed hatchet cell in the wings, as found in tsetse-flies.

9.4.2   Life cycle

Both males and females take blood-meals from wild and domesticated animals, including cattle, horses, pigs and dogs; they also feed on humans, especially if their preferred hosts are absent or scarce in the area. Their bites can be painful and most are on the legs. During feeding the forward-projecting proboscis is swung downwards, and the skin penetrated. In hot weather, flies digest their blood-meals within 12–24 hours and feed about every 1–3 days, but in cooler conditions blood digestion is prolonged to 2–4

Figure 9.7 Lateral view of the head of *Stomoxys calcitrans* showing the forward-projecting proboscis.

days or more, and feeding occurs every 5–10 days. Biting is restricted to the daylight hours and occurs both in bright sunshine and in cloudy overcast weather. Most biting occurs outdoors but stable-flies will also enter houses to feed. They are mostly encountered in and around farms or where horses are kept, and consequently are more common in rural areas than in towns.

The creamy-white eggs, 1 mm long, resemble those of house-flies. They are normally laid in batches of less than 20, but sometimes as many as 50–200 may be laid together. Eggs are usually deposited in horse manure but also in compost pits, decaying and fermenting piles of vegetable matter, weeds, cut grass or hay. Stable-flies very rarely lay their eggs in human or animal faeces, unless these are liberally mixed with hay or straw.

Eggs hatch within 1–5 days. The resultant larvae are creamy-coloured maggots which resemble those of the house-fly. However, they can be separated by the arrangement of the posterior spiracles, which are widely separated (Fig. 9.6a), thus differing from the more closely positioned spiracular plates of *Musca* and *Muscina*. They are also approximately round in outline, lack a conspicuous peritreme, and the S-shaped spiracular slits are widely separated from each other. Larvae prefer a high degree of moisture for development and therefore are found mostly in wet mixtures of manure and soil or straw, and in vegetable matter in advanced stages of decay. Under optimum conditions the larval period lasts about 6–10 days, but in cooler weather, or when there is a shortage of food, larval development can be prolonged to 4–5 weeks or more.

Larvae migrate to drier areas and bury themselves in the soil prior to pupation. The puparium is dark brown and resembles that of the house-fly, but can be distinguished from it by having the posterior spiracles widely separated. The puparial stage usually lasts 5–7 days, although in cool weather it may be extended to several weeks. The life cycle from egg laying to adult emergence may last from 12 to 42 days, the time depending mainly on temperature.

In tropical areas stable-flies breed continuously throughout the year, but in more temperate climates they pass the cooler months as larvae or puparia. Sometimes adults survive the winter in warm stables or buildings, feeding intermittently during the cooler months.

9.4.3   Medical importance

Because of the painful bites inflicted by both sexes stable-flies can be troublesome pests of people, cattle, horses and pets – notably dogs. Adult flies are not regarded as transmitting diseases to humans, and although they have been considered as mechanical vectors of African trypanosomiasis there is little evidence that they play any part in the epidemiology of sleeping sickness. However, they do appear to transmit *Trypanosoma evansi*, the causative agent of Surra, a disease which infects camels, horses and other animals in many tropical countries. Because stable-flies rarely visit excreta and festering wounds they are much less likely to spread pathogens in the way that house-flies do.

In the tropical Americas eggs of *Dermatobia hominis*, a myiasis-producing fly (Chapter 10), are sometimes attached to adult stable-flies.

9.4.4   Control

Many of the control methods aimed at house-flies can be applied with some modification to control stable-flies – for example, not allowing piles of manure, grass cuttings or decaying vegetable matter to accumulate, and burning straw and bedding material. Breeding places can be sprayed with organochlorine or organophosphate insecticides, but as noted in the section on house-fly control it is difficult to get deep insecticidal penetration into the breeding places where most larvae are found. Insecticidal spraying of horse stables, animal shelters, barns and other farm buildings can help reduce their numbers.

9.5   FANNIIDAE: LESSER HOUSE-FLIES AND LATRINE FLIES (*FANNIA* SPECIES)

9.5.1   External morphology

Flies of the genus *Fannia* resemble house-flies but are rather smaller (6–7 mm), and are readily distinguished from house-flies and other similar flies by their wing venation and antennae (Fig. 9.3). In *Fannia* vein 4 of the wing is more or less parallel to vein 3, whereas in *Musca* it bends upwards and almost touches vein 3 at the wing apex, and in *Muscina* vein 4 also bends upwards but not to such an extent. The arista, which arises from the third antennal segment, is completely devoid of hairs (Fig. 9.3), whereas in both *Musca* and *Muscina* it bears branches on both sides. In most other respects the adult flies are similar to house-flies. There are two common species of *Fannia* which are of minor medical importance, namely *Fannia canicularis* (the lesser house-fly), which occurs world-wide and is commonly encountered in houses, and *Fannia scalaris* (the latrine fly), which has a Holarctic distribution and is less common in houses. *Fannia canicularis* has three longitudinal stripes on the thorax, whereas *F. scalaris* has two such stripes.

9.5.2   Life cycle

The lesser house-fly (*F. canicularis*) lays about 50–100 eggs, which resemble those of the common house-fly. They are deposited on people's food, but also on urine-soaked bedding of humans and animals, compost heaps, decaying piles of grass, human and animal excreta and in poultry litter. The latrine fly (*F. scalaris*) usually lays her eggs in faeces – hence the common name. Larvae of *Fannia* species prefer wetter breeding places to

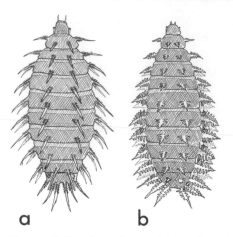

Figure 9.8 Final-instar larvae. (*a*) *Fannia canicularis*; (*b*) *Fannia scalaris*.

house-flies and are well adapted to living in liquid habitats, such as semi-liquid faeces.

Eggs hatch after 1–2 days. *Fannia* larvae are quite distinct from the maggot-shaped larvae of *Musca* and are unlikely to be confused with the larvae of any other medically important fly (Fig. 9.8). They are flattened dorsoventrally and have many thin but conspicuous fleshy processes arising from the body segments which bear small spiniform secondary processes. Larval development takes about 7–12 days but may be prolonged if the habitat starts to dry out. The puparium is brown in colour and is similar to the shape of the larva. After 7–10 days the adult fly emerges. The life cycle often lasts about 1 month, which is considerably longer than in *Musca domestica*, but may under favourable conditions be completed within 13–22 days. The feeding mechanism of the adult is similar to that of *M. domestica*. Although *Fannia* adults often enter houses they do not settle on people, or on their food, as much as house-flies.

9.5.3   Medical importance
Many of the pathogens transmitted by the house-fly are probably also spread by *Fannia* species. They have been incriminated in cases of urogenital myiasis, and larvae are sometimes found in stools, but as previously stressed true intestinal myiasis does not occur in humans.

9.5.4   Control
The same control methods apply to species of *Fannia* as to *Musca*, but particular attention should be given to the prevention and eradication of *Fannia scalaris* breeding in latrines.

FURTHER READING

Bidawid, S.P., Edeson, J.F.B., Ibrahim, J. and Matossian, R.R. (1978) The role of non-biting flies in the transmission of enteric pathogens (*Salmonella* species and *Shigella* species) in Beirut, Lebanon. *Annals of Tropical Medicine and Parasitology*, **72**, 117–21.

Chavasse, D.C., Shier, R.P., Murphy, O.A., Huttly, S.R.A., Cousens, S.N. and Akhtar, T. (1999) Impact of fly control on childhood diarrhoea in Pakistan: community-randomised trial. *Lancet*, **353**, 22–5.

Cohen, D., Green, M., Block, C., Slepon, R., Ambar, R., Wasserman, S.S. and Levine, M.M. (1991) Reduction of transmission of shigellosis by control of houseflies (*Musca domestica*). *Lancet*, **337**, 993–7.

Curtis, C. (1998) The medical importance of domestic flies and their control. *Africa Health*, **20** (6), 14–15.

Echeverra, P., Harrison, B.A., Tirapat, C. and McFarland, A. (1983) Flies as a source of enteric pathogens in a rural village in Thailand. *Applied and Environmental Microbiology*, **46**, 32–6.

Emerson, P.M., Lindsay, S.W., Walraven, G.E.L., Faal, H., Bøgh, C., Lowe, K. and Bailey, R.L. (1999) Effect of fly control on trachoma and diarrhoea. *Lancet*, **353**, 1401–3.

Feacham, R.G., Bradley, D.J., Garelick, H. and Duncan Mara, D. (eds.) (1983) *Sanitation and Disease: Health Aspects of Excreta and Water Management*. Chichester: Wiley.

Greenberg, B. (1971) *Flies and Disease*, vol. 1, *Ecology, Classification and Biotic Associations*. Princeton, New Jersey: Princeton University Press.

Greenberg, B. (1973) *Flies and Disease*, vol. 2, *Biology and Disease Transmission*. Princeton, New Jersey: Princeton University Press.

Keiding, J. (1986) *The House-Fly: Biology and Control*. WHO/VBC/86.937. Geneva: World Health Organization mimeographed document.

Lindsay, D.R., Stewart, W.H. and Watt, J. (1953) Effect of fly control on diarrheal diseases in an area of moderate morbidity. *Public Health Reports, Washington, DC*, **68**, 361–7.

O'Hara, J.E. and Kennedy, M.J. (1991) Development of the nematode eyeworm, *Thelazia skrjabini* (Nematoda: Thelazioidea), in experimentally infected face flies, *Musca autumnalis* (Diptera: Muscidae). *Journal of Parasitology*, **77**, 417–25.

Skidmore, P. (1985) *The Biology of the Muscidae of the World*. Dordrecht: W. Junk.

Sulaiman, S., Sohadi, A.R., Yurms, H. and Iberahim, R. (1988) The role of some cyclorrhaphan flies as carriers of human helminths in Malaysia. *Medical and Veterinary Entomology*, **5**, 72–80.

West, L.S. (1951) *The House-fly: Its Natural History, Medical Importance and Control*. Ithaca, New York: Comstock Publishing/Cornell University Press.

Zumpt, F. (1973) *The Stomoxyine Biting Flies of the World. Diptera: Muscidae. Taxonomy, Biology, Economic Importance and Control Measures*. Stuttgart: Gustav Fischer.

10

Flies and myiasis

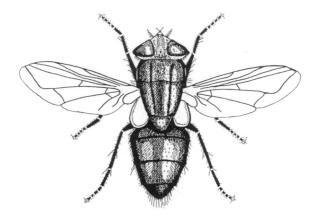

Myiasis can be defined as the invasion of organs and tissues of humans or other vertebrate animals with dipterous larvae, which for at least a period feed upon the living or dead tissues or, in the case of intestinal myiasis, on the host's ingested food. Different terms can be used to describe myiasis which affects different parts of the body – for example, cutaneous, dermal or subdermal myiasis; urogenital myiasis; ophthalmic myiasis; nasopharyngeal myiasis; and intestinal or enteric myiasis. When larvae burrow just under the surface layers of the skin this is sometimes called creeping eruption or creeping myiasis; when boil-like lesions are produced the term furuncular myiasis may be used, and when wounds become infested this is often referred to as traumatic myiasis.

Myiasis may be obligatory or facultative. In obligatory myiasis it is essential for the fly larvae to live on a live host for at least a certain part of their life. For example, larvae of *Cordylobia anthropophaga*, *Cochliomyia hominivorax*, *Chrysomya bezziana*, *Dermatobia hominis* and *Wohlfahrtia magnifica* are obligatory parasites of humans and other vertebrates. In contrast, in facultative myiasis larvae are normally free-living, often attacking carcasses, but under certain conditions they may infect living hosts. Several types of fly, including species of *Calliphora*, *Lucilia*, *Phormia* and *Sarcophaga*, which normally breed in meat or carrion, may cause facultative cutaneous myiasis in people by infecting festering sores and wounds.

There is no obligatory intestinal myiasis of humans. When people have maggots in their intestinal tract this is most likely due to accidental swallowing of eggs or larvae on food. Although the maggots may be able to survive for some time in the intestine, there is no species of fly which is specially adapted to cause intestinal myiasis in humans. In contrast, obligatory intestinal myiasis occurs in animals. The presence of larvae in the human intestine may nevertheless cause considerable discomfort, abdominal pain and diarrhoea, which may be accompanied by discharge of blood and vomiting. Living larvae may be passed with the excreta or vomit. Occasionally facultative urogenital myiasis occurs in humans; this usually involves larvae of *Musca* or *Fannia* species. It seems that ovipositing flies are attracted to unhygienic discharges and lay their eggs near genital orifices. When these eggs hatch the minute larvae enter the genital orifice and work their way up the urogenital tract. Much pain may be caused by larvae obstructing these passages, and mucus and blood, and eventually larvae, may be discharged with the urine.

When larvae occur in wounds, sores and dermal or subdermal tissues, their removal under aseptic conditions is usually a relatively simple procedure. When they are more deeply imbedded in the underlying tissues, or when they have penetrated the mucous membranes, eyes, frontal sinuses or cavities, their removal is more difficult and surgery may be needed. Major and irreversible damage may have been done by the larvae.

The biologies and medical importance of the principal types of flies causing facultative and obligatory myiasis in humans are outlined below.

10.1 CLASSIFICATION

The flies in this chapter are contained in three families: the Calliphoridae, the Sarcophagidae and the Oestridae. The family Calliphoridae can be conveniently divided into a group containing the non-metallic flies such as the Congo floor-maggot fly (*Auchmeromyia senegalensis* (=*luteola*)) and the tumbu fly (*Cordylobia anthropophaga*), and another group containing the metallic calliphorids. Well-known examples of the metallic group are the blowflies, comprising the bluebottles (*Calliphora*) and the greenbottles (*Lucilia*), the New World screwworms (*Cochliomyia hominivorax*) and the Old World screwworms (*Chrysomya bezziana*).

10.2 CALLIPHORIDAE: NON-METALLIC FLIES

10.2.1 *Cordylobia anthropophaga*

External morphology

Cordylobia anthropophaga is known as the tumbu or mango fly and is found in Africa, occurring from Ethiopia in the north, through West and East Africa to Natal and the Transvaal in the south. It is not found outside Africa except possibly in southwestern Saudi Arabia.

Adults are robust, relatively large flies, 9–12 mm long, dull yellowish to light brown in colour, but with two dark grey and poorly defined dorsal longitudinal thoracic stripes (Fig. 10.1). There are four visible abdominal segments which are more or less equal in length (compare *Auchmeromyia senegalensis* in which the second abdominal segment is markedly longer than the others). The wings are slightly brownish.

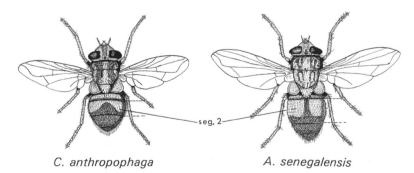

C. *anthropophaga*   A. *senegalensis*

Figure 10.1 Adults of the tumbu fly (*Cordylobia anthropophaga*) and the Congo floor-maggot fly (*Auchmeromyia senegalensis*) showing the longer second abdominal segment in *A. senegalensis*.

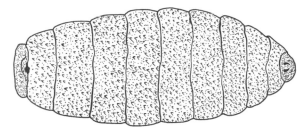

Figure 10.2 Final-instar larva of the tumbu fly (*Cordylobia anthropophaga*).

Life cycle

Females lay up to 100–300 eggs in batches on dry soil and sand in shady places, especially those contaminated with the urine or excreta of humans, rodents, dogs or monkeys. Females also oviposit on underclothes or soiled babies' nappies (diapers) placed on the ground to dry. The eggs, which are white and banana-shaped, hatch after about 1–3 days. Larvae attach themselves to a suitable host either directly, or attach themselves temporarily to washed clothing placed on the ground to dry, and so get transferred to people if the clothing is not ironed before it is worn. Once on a host, powerful hook-like mouthparts enable a larva to bury itself completely except for its posterior spiracles situated at the tip of the abdomen, which remain in contact with the air. Newly emerged larvae can live as long as 9–15 days on the ground in the absence of a suitable host before they die.

The minute first-instar larvae are typically maggot-shaped (like e.g. the larvae of house-flies: see Fig. 9.4*b*), the second-instar larvae are club-shaped, and the third and final instar larvae are rather fat, broadly oval-shaped, yellowish-white maggots, about 11–15 mm long. They are covered with numerous spicules which are often, but not always, grouped into three or more transverse rows per segment (Fig. 10.2). After 10–12 days mature larvae wriggle out of the boil-like swellings and fall to the ground where they bury themselves and turn into puparia. Adult flies emerge 8–15 days later and readily enter houses, where they may lay their eggs on the mud floors, especially if children have urinated on them.

Medical importance

Larvae of *Cordylobia* cause boil-like (furuncular) swellings on almost any part of the body. Although these swellings may become sore and inflamed, and even quite hard and exude serous fluids, they do not usually contain pus. Generally just one or two larvae are found in a patient; however, more than 60 larvae have been recovered from a person, mostly in separate lesions. The standard method of extracting a larva is to cover the small hole in the swelling with medicinal liquid paraffin. This prevents the larva from breathing through its posterior spiracles with the result that it wriggles a little further out of the swelling to protrude the spiracles. In so doing it lubricates the pocket in the skin, and the larva can then usually be extracted by gently pressing around the swelling.

Infections can be prevented, or at least minimized, by ensuring that clothes, bed linen and towels are not spread on the ground to dry, and also by ironing clothing.

Dogs and rats are commonly infected with tumbu larvae.

10.2.2 *Auchmeromyia senegalensis*

External morphology
Auchmeromyia senegalensis, commonly known as the Congo floor-maggot fly, although not strictly speaking producing myiasis, is described here because the adults are often confused with those of *Cordylobia anthropophaga*. It occurs throughout Africa south of the Sahara and also in the Cape Verde islands.

Adults are very similar to *C. anthropophaga* but are distinguished by the second abdominal segment being about twice as long as any of the others (Fig. 10.1), whereas in the tumbu fly all segments are about equal in length.

Life cycle and medical importance
Eggs are laid in batches of about 50 on the dry sandy floors of mud huts. They hatch after 1–3 days and the larvae hide away in cracks and crevices in the hut floor, especially under beds and sleeping-mats. At night the larvae crawl out from these daytime refuges and take blood-meals from people sleeping on the floor. After taking a full blood-meal the now pinkish larvae return to their hiding places. Larvae may feed four or five times a week, but can withstand starvation for long periods in the absence of suitable hosts. There are three larval instars each requiring at least two blood-meals. Under optimum conditions larval development is completed within 3–4 weeks, but this period may be prolonged to as long as 3 months if the larvae fail to obtain regular feeds. The fully developed third-instar larvae, unlike those of *Cordylobia anthropophaga*, are not covered with spicules (Fig. 10.3).

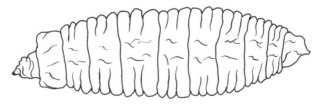

Figure 10.3 Final-instar larva of the Congo floor-maggot fly (*Auchmeromyia senegalensis*).

Larvae cannot climb, and therefore people will not be attacked if they sleep on beds raised from the floors by only short legs. Mature larvae pupate in cracks or directly on the surface of the mud floor of huts. Adults emerge from the puparia after about 9–16 days.

The Congo floor-maggot fly used to be quite common in certain parts of Africa, but due to the changes in life-style of most Africans it is becoming increasingly rare, and these days is not much of a problem.

10.3   CALLIPHORIDAE: METALLIC FLIES

10.3.1 New World screwworms

The New World screwworm, *Cochliomyia hominivorax*, used to occur in the southern states of the USA, through Mexico, Central America, the Caribbean islands, into the northern parts of South America. However, following successful eradication programmes, achieved by the release of thousands of sterilized male flies, the screwworm has been eradicated from the USA, Mexico and parts of Central America. An eradication programme is presently continuing in Central America to push the fly's northern limits to Panama. In 1988 a population of the New World screwworm was discovered in Libya, but was eradicated in 1991.

External morphology

Adult flies are 8–10 mm long, metallic green to bluish-green in colour and have three distinct dark longitudinal stripes on the dorsal surface of the thorax (Fig. 10.4). The dorsal bristles on the thorax, like those of *Chrysomya*, are poorly developed, thus distinguishing screwworm flies from *Lucilia* and *Calliphora*, which have well-developed bristles. The thoracic squama is

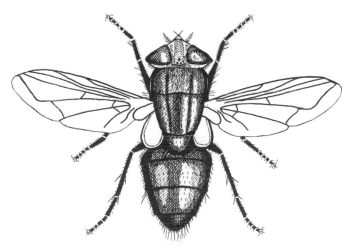

Figure 10.4 Adult of a New World screwworm (*Cochliomyia hominivorax*).

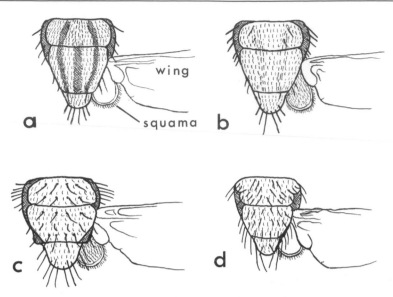

Figure 10.5 Thoraces and bases of right wing showing the presence or absence of prominent thoracic dorsal bristles, and the presence or absence of fine hairs on thoracic squama of the wing. (a) *Cochliomyia*, note the three dark thoracic stripes and squama without hairs; (b) *Chrysomya*, note the absence of prominent thoracic stripes and hairy squama; (c) *Calliphora*, note the prominent thoracic bristles and hairy squama; (d) *Lucilia*, note the prominent thoracic bristles and hairless squama.

a membranous lobe on the posterior border of the wing near the thorax and in New World screwworms (Fig. 10.5a) is similar to that of *Lucilia* in being devoid of hairs.

Life cycle

Females of *C. hominivorax* lay batches of 10–490 eggs on the edges of wounds, scabs, sores, scratches or pimples, also on dried blood clots and on diseased and even healthy mucous membranes such as nasal passages, eyes, ears, the mouth and vagina. In newborn babies eggs can be laid in the umbilicus. Eggs hatch within 11–24 hours and the active larvae burrow deeply into the living tissues and feed gregariously. There are three larval instars and the third-instar larvae, which are formed after 2–3 days, are about 15–17 mm long and typically maggot-shaped. They are distinguished from house-fly maggots by the presence of distinct bands of spicules encircling the anterior margins of all body segments (Fig. 10.6a). Larvae tend to penetrate deeply into tissues, so that infections near the eyes, nose and mouth can cause considerable destruction of these areas, often accompanied by putrid-smelling discharges and ulcerations.

After 4–8 days the larvae reach maturity and wriggle out of the wounds or passages they have excavated by devouring living tissue and drop to the ground, where they bury themselves in the soil and pupate. In warm weather the puparial stage lasts about 7–10 days, but in cooler weather may be prolonged for many weeks or even months. Under optimum conditions the life cycle from egg to adult is 11–14 days.

10.3.2 Old World screwworms

The genus *Chrysomya* contains many species and is common in the tropics. About 10 species are known to cause myiasis in humans, but only *Chrysomya bezziana* is important because its larvae are obligatory parasites of living tissues. Larvae of the other species are not obligatory parasites and often develop in carrion and decomposing matter. *Chrysomya bezziana* occurs throughout tropical Africa, the Indian subcontinent, throughout most of South-east Asia to China, and the Philippines to Papua New Guinea; it has been introduced to several countries on the west coast of the Persian Gulf, including Iraq. Other less important facultative myiasis-producing species such as *C. albiceps* and *C. megacephala* of the Old World have invaded Central and South America.

External morphology

Adults of *Chrysomya bezziana* are similar to those of *Cochliomyia hominivorax*, but they lack the distinctive longitudinal thoracic stripes (Fig. 10.5*b*), and the dorsal surface of the thoracic squama is covered with fine hairs (Fig. 10.5*b*). Larvae are very similar to those of *C. hominivorax*, although they differ in usually having four to six finger-like processes on the anterior

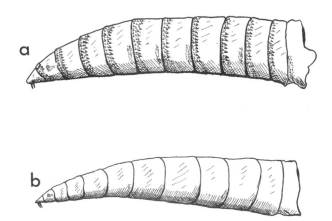

Figure 10.6 Final-instar larvae. (*a*) *Cochliomyia hominivorax*; (*b*) *Lucilia* species (*Calliphora* larvae are almost identical).

spiracles, not the usual seven to nine found in *C. hominivorax*. In practice separation is easy: the screwworms of the Americas belong to the genus *Cochliomyia* whereas those of the Old World belong to the genus *Chrysomya*.

Life cycle

The life cycle of *Chrysomya bezziana* is very similar to that of *Cochliomyia hominivorax*. About 150–500 eggs are deposited in wounds, open sores, scabs, ulcers, scratches or on mucous membranes, especially those contaminated with discharges. Newly emerged larvae burrow through the skin to the underlying tissues where they commonly remain congregated together. Larvae complete their development in 5–6 days and then wriggle out of the wounds and drop to the ground, where they bury themselves and pupate. The puparial period lasts about 7–9 days in warm weather, but is prolonged to several weeks or even months during cold weather. The life cycle from egg to adult under ideal conditions is about 12–14 days.

Adults of both the New and Old World screwworms are frequently found feeding on decomposing corpses, decaying matter, excreta and flowers.

Medical importance of screwworms

Larvae of both *Chrysomya bezziana* and *Cochliomyia hominivorax* are obligatory parasites of living tissues and cause human myiasis, which can be very severe resulting in considerable damage and disfigurement, especially if the face is attacked. When larvae invade natural orifices, such as the nose, mouth, eyes or vagina, they can cause excruciating pain and misery. In one patient suffering from a nasal infection 385 larvae of *C. hominivorax* were removed over a 9 day period! Larvae of both species may eat their way through the palate and as a result impair speech.

Chrysomya bezziana causes more cases of myiasis in people living in India and other parts of Asia than it does in Africa. *Chrysomya* adults often visit faeces, so are potential vectors of several pathogens.

All myiasis cases should be treated immediately because the very rapid larval development can cause permanent damage. Formerly maggots in open wounds or body openings were often removed by irrigating infested areas with 5–15% chloroform in light vegetable oil, but because of health and safety regulations, ethanol can be substituted for chloroform. Surgery may be necessary to expose deeply embedded larvae. There are serious problems when larvae are in the eye orbit and delicate surgery may be needed to remove them; permanent damage may have been done.

Both screwworm species, but particularly *C. hominivorax*, cause myiasis in cattle, goats, sheep and horses and are responsible for enormous economic losses to the livestock industry. The technique of releasing millions of laboratory-reared sterile male flies into areas infested with screwworms has given good control, and has even resulted in the eradication of *C. hominivorax* from the USA, Mexico, Guatemala, Belize, Honduras and El Salvador, and some Caribbean islands, as well as from Libya.

10.3.3 The greenbottles (*Lucilia*)

External morphology
There are several species of greenbottles within the genus *Lucilia* and although the genus has a world-wide distribution most species occur in northern temperate regions. They are mostly metallic or coppery green in colour, usually a little smaller (about 10 mm long) and a little less bristly than species of *Calliphora* (bluebottles). As in *Calliphora*, prominent bristles are present on the dorsal surface of the thorax (Fig. 10.5*d*), but whereas the squama of the wing is hairy dorsally in bluebottles it is without hairs in *Lucilia* (Figure 10.5*d*). *Lucilia sericata* is the commonest species, occurring in the Americas, Europe, Asia, Africa, Australia and in most other areas of the world. Another common species, *L. cuprina*, occurs mainly in Africa, Asia and Australia.

Life cycle
Female greenbottles normally lay their eggs on meat, fish, carrion and decaying or decomposing carcasses, but they will also oviposit on or near festering and foul-smelling wounds of humans and animals, and on excreta and decaying vegetable matter. Eggs hatch within 8–12 hours. Larvae are typically maggot-shaped (Fig. 10.6*b*). They may bear small spicules but not conspicuous ones as do larvae of screwworms, which they otherwise tend to resemble. They can be distinguished from screwworms by the shape of the slits in the posterior spiracles.

The larval period lasts about 4–8 days. Mature larvae bury themselves in loose soil and pupate; the puparial period lasts about 6–14 days.

Adult flies frequently visit carrion, excreta, general refuse, decaying material, sores and wounds. They are particularly common around unhygienic places and situations where meat or decaying animals are present. They are nearly always abundant near slaughterhouses and piggeries. They commonly fly into houses, where they are particularly troublesome because of their noisy buzzing flight. The most common species infesting wounds of humans are *Lucilia sericata* and *L. cuprina*.

10.3.4 The bluebottles (*Calliphora*)

External morphology
The flies usually known as bluebottles belong to the genus *Calliphora*. Although *Calliphora* has a world-wide distribution bluebottles are more

common in the northern temperate regions than in the tropical or southern temperate regions.

Adults are robust flies, mostly dull metallic-bluish or bluish-black in colour and 8–14 mm long. As in *Lucilia* there are well-developed bristles on the thorax (Fig. 10.5c), but the squama of the wing is hairy on the dorsal surface (Fig. 10.5c) whereas in *Lucilia* it lacks hairs. The abdomen is rather more shiny than the thorax.

Life cycle
Bluebottle larvae look very similar to those of *Lucilia*, and the life cycle is also very similar to that described for *Lucilia*.

Medical importance of greenbottles and bluebottles
The dirty habit of blowflies (greenbottles and bluebottles) of alighting and feeding on excreta, decaying material and virtually all common foods of humans makes them potential vectors of a number of pathogens. However, their medical importance is usually associated with facultative myiasis.

Larvae of both *Lucilia* and *Calliphora* have been found in many parts of the world developing in foul-smelling wounds and ulcerations, especially those producing pus. They have also been recorded in hospitals underneath the bandages and dressings of patients, especially when these have become contaminated with blood and pus. Such infections do not usually cause any serious damage or harm since the larvae feed mainly on pus and dead tissues. During the Napoleonic wars (1796–1814) in Europe surgeons used maggots to cleanse septic battle wounds, and this practice was continued in some parts of the world until about the 1950s. Recently maggot therapy has been rediscovered and some doctors, particularly in the USA and Britain, are advocating the use of sterile larvae to clean up leg ulcers, pressure sores and infected wounds. Some patients need considerable persuasion to accept this unusual form of treatment!

Removal of maggots of *Lucilia* and *Calliphora* usually presents no problems as they can be picked out of wounds with sterile forceps and antibiotic dressings applied. Very rarely maggots which have infected wounds invade healthy tissues. Occasionally intestinal myiasis is reported. This is usually caused by eating uncooked foods contaminated with larvae of *Lucilia* or *Calliphora*, but usually the larvae are killed within the human alimentary canal and no serious harm is done. As emphasized previously (p. 151) there is no obligatory intestinal myiasis in humans.

Control of greenbottles and bluebottles
The principal breeding places of greenbottles and bluebottles are domestic refuse, rubbish tips, dustbins, offal and other wastes of slaughterhouses and meat packing factories. Larvae also occur in foods such as meat and

fish left out in the sun to dry. Any methods which reduce the accumulation of these potential breeding sites are welcome. Dustbins and garbage cans should have tight-fitting lids and be emptied once or twice a week. The outside of such bins and both sides of the lid can be sprayed with orga-nophosphate or carbamate insecticides every 7–10 days, and the adjacent walls and fences sprayed every 2 weeks. This will deter flies from ovipos it-ing in these sites.

The reader is referred to Chapter 9 for control measures against house-flies and related flies, because many of the insecticides and methods used are applicable to controlling the Calliphoridae.

10.4   SARCOPHAGIDAE

Only the genera *Sarcophaga* and *Wohlfahrtia* of the Sarcophagidae, both of which have several species, are of any medical importance. They cause myiasis, and possibly act as mechanical vectors of pathogens. They are sometimes called flesh-flies. They are unusual in that females are larvipar-ous, that is they deposit first-instar larvae instead of laying eggs. They have a world-wide distribution.

10.4.1 *Sarcophaga*

External morphology
Sarcophaga species are large and hairy non-metallic flies, about 10–15 mm long, and usually greyish in colour. They have three prominent black longi-tudinal dorsal stripes on the thorax. The abdomen is sometimes distinctly, but other times indistinctly, marked with squarish dark patches on a grey background giving it a chequer-board (chess-board) appearance (Fig. 10.7).

Life cycles and medical importance
Adult *Sarcophaga* do not lay eggs but deposit first-instar larvae, as do tsetse-flies and *Wohlfahrtia* species. The larvae are deposited in batches of

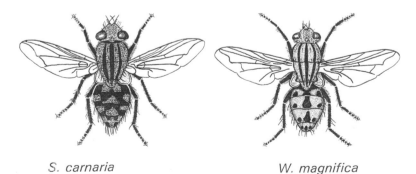

S. carnaria   W. magnifica

Figure 10.7 Adults of *Sarcophaga carnaria* and *Wohlfahrtia magnifica* showing differences in abdominal markings.

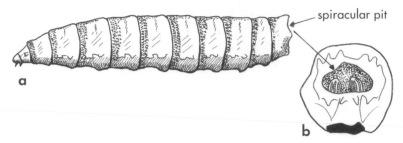

spiracular pit

a

b

Figure 10.8 Larvae of *Sarcophaga carnaria* showing the deep pit in which the posterior spiracles are located (*Wohlfahrtia* larvae are almost identical).

40–60, usually on decaying carcasses, rotting food and human and animal excreta, but sometimes in wounds. They are primarily scavengers. Larvae are typically maggot-shaped. They are distinguished from larvae of the Calliphoridae by the posterior spiracles being situated in a deep pit (Fig. 10.8), and thus difficult to see, and also by having bands of spicules on the body. Larvae of *Sarcophaga* are virtually indistinguishable from those of *Wohlfahrtia* species.

Larval development is rapid, lasting in hot weather in the presence of a nutritious food supply only 3–4 days, at the end of which time the larvae bury themselves in the soil and pupate. The puparial stage lasts about 7–12 days.

Although larvae are normally deposited in carrion they very occasionally occur in wounds, but usually they cause little damage as they feed mainly on necrotic tissues. They have more commonly been incriminated in accidental intestinal myiasis, causing considerable discomfort and pain before the larvae are passed out with the faeces. The most common species is *Sarcophaga cruentata* (=*haemorrhoidalis*), which is widely distributed in the Americas, Europe, Africa and Asia. Because adults frequent festering wounds, excreta and decaying animal matter they may be mechanical vectors of various pathogens.

10.4.2 *Wohlfahrtia*

External morphology
Wohlfahrtia species are hairy flies about as large as, or a little larger than, bluebottles. They are greyish and like *Sarcophaga* have three distinct black lines on the dorsal surface of the thorax. The dark markings on the abdomen, however, are not in the form of a chess-board pattern as in *Sarcophaga* species, but are usually present as roundish lateral spots and triangular-shaped dark markings along the midline (Fig. 10.7). There is, though, considerable variation and sometimes the dark marks are so

large as to be more or less confluent, making the abdomen appear mainly black.

Life cycle and medical importance
As with *Sarcophaga* and tsetse-flies, adults of *Wohlfahrtia* deposit larvae not eggs. The most important species is *W. magnifica*, an obligate parasite of humans and animals (e.g. camels, domestic livestock and dogs) in western Europe, the Middle East, North Africa and Central Asia to China. Some 120–170 larvae are deposited, often in several batches, in scratches, wounds, sores and ulcerations. In people the ears, eyes and nose are frequently infested, and this can result in deafness, blindness and even death. The North American species, *W. vigil*, will also deposit its larvae on unbroken skin if it is soft and tender, such as in babies and very young children, who are, therefore, more commonly attacked than adults. *Wohlfahrtia* larvae have bands of spicules, and the spiracles are situated in a deep pit (Fig. 10.8). They are very similar in appearance to those of *Sarcophaga*. Larval development takes 5–9 days, after which mature larvae drop to the ground, bury themselves in loose soil and then pupate. Adults emerge from the puparia after 8–12 days.

Like the larvae of screwworms, *Wohlfahrtia* larvae can burrow deep into the tissues and cause considerable damage.

10.5   OESTRIDAE
The family Oestridae comprises four subfamilies, three of which (Oestrinae, Gasterophilinae and Hypodermatinae) contain important obligate parasites of domestic animals. The subfamily Cuterebrinae contains several species in six genera that cause myiasis in rodents, monkeys and livestock, and also *Dermatobia hominis*. This is the so-called human bot-fly which causes obligatory myiasis in people and animals living in Central and South America.

10.5.1 *Dermatobia hominis*

External morphology
Adults of *Dermatobia hominis* are a little larger (12–18 mm) than bluebottles (*Calliphora*) but have a similar dark-blue metallic-coloured abdomen, dark bluish-grey thorax and a mainly yellowish head (Fig. 10.9). The mouthparts are vestigial.

Life cycle
Dermatobia hominis occurs primarily in lowland forests, being especially common along woodland paths and at the margins of forest and scrub areas. These flies have an interesting and remarkable life history.

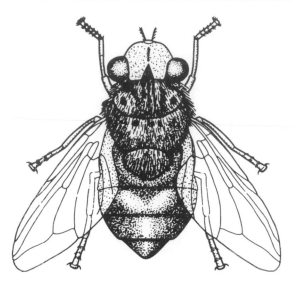

Figure 10.9 Adult of *Dermatobia hominis*.

Females glue 6–30 eggs to the body of other arthropods, such as mosquitoes (especially of the genus *Psorophora*), house-flies, stable-flies and even ticks.

Embryos within the attached eggs mature into first-instar larvae within 4–9 days. They do not hatch, however, until the insects carrying the eggs settle on humans or some other warm-blooded animal, even birds, to take a blood-meal or, as in the case of house-flies, feed on sweat. The larvae then emerge from the eggs, which remain attached to the insect carrier, and drop on to the host's skin. Here, within 5–10 minutes, the larvae manage to penetrate the skin and burrow into the subcutaneous tissues. Each larva produces a boil-like swelling which has an opening through which the larva breathes. Occasionally two or three larvae may be found within a single swelling.

The first-instar larvae are 1–1.5 mm long, more or less cylindrical in shape (Fig. 10.10*a*), and have the anterior half of the body covered with numerous spines of two different sizes. The second-instar larvae are a completely different shape, being enlarged anteriorly but with the posterior half of the body distinctly narrow, giving the appearance of a bottle with a long neck. Relatively large thorn-like spines encircle the middle segments (Fig. 10.10*b*). The third and final instar larvae are about 18–25 mm long, more or less oval and have relatively small spines on the anterior segments (Fig. 10.10*c*). A pair of very distinct flower-like spiracles are present anteriorly, and less conspicuous slit-like spiracles are situated in a small concavity of the last abdominal segment.

Larval development is completed in a small pocket excavated in the

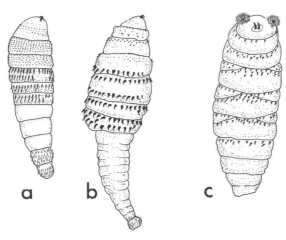

Figure 10.10 *Dermatobia hominis* larvae. (*a*) First-instar; (*b*) second-instar; (*c*) third- and final-instar.

subdermal layer of the host, and lasts about 4–18 weeks. Mature larvae wriggle out of the skin and drop to the ground where they pupate just under the surface of the soil. Adult flies emerge from the puparia after about 5–12 weeks, but are rarely seen.

Medical importance
Larvae of *Dermatobia hominis* invade the subcutaneous tissues of humans on various parts of the body, including the head, arms, abdomen, buttocks, thighs, scrotum and axillae. They produce boil-like swellings which suppurate and this may attract other myiasis-producing flies to the host. They can cause a lot of discomfort and considerable pain. Because of the long duration of the larval life (up to 18 weeks) infected persons may be encountered in almost any part of the world following air travel.

Because of the many spines on the larvae and their shape it is often difficult to remove them by squeezing them out, a procedure that works well with the African tumbu fly; consequently surgical removal of larvae under sterile conditions may be necessary. Frequently a local anaesthetic is needed.

10.6   OTHER MYIASIS-PRODUCING FLIES
Several other species of flies commonly cause myiasis in livestock. There are, for example, flies infesting donkeys and camels, others infesting sheep and goats (e.g. *Oestrus ovis*), cattle (e.g. *Hypoderma bovis*) and horses and donkeys (e.g. *Gasterophilus haemorrhoidalis*). Occasionally humans become infected. Those most likely to suffer are people working with the flies' natural hosts; for example, shepherds looking after sheep may become infected with *O. ovis*.

FURTHER READING

Abram, L.J. and Froimson, A.I. (1987) Myiasis (maggot infection) as a complication of fracture management: a case report and review of the literature. *Orthopedics*, **10**, 625–7.

Arbit, E., Varon, R.E. and Brem, S.S. (1986) Myiatic scalp and skull infection with Diptera *Sarcophaga*: a case report. *Neurosurgery*, **18**, 361–2.

Bunkis, J., Gherini, S. and Walton, R.L. (1985) Maggot therapy revisited. *Western Journal of Medicine*, **142**, 554–6.

Catts, E.P. (1982) Biology of New World bot flies: Cuterebridae. *Annual Review of Entomology*, **27**, 313–38.

Davis, E. and Shuman, C. (1982) Cutaneous myiasis: devils in the flesh. *Hospital Practice*, 1982 (December), 115–23.

Gabaj, M.M., Gusbi, A.M. and Awan, M.A.Q. (1989) First human infestations in Africa with larvae of the American screw-worm, *Cochliomyia hominivorax* Coq. *Annals of Tropical Medicine and Parasitology*, **83**, 553–4.

Graham, O.H. (ed.) (1985) Symposium on eradication of the screw-worm from the United States and Mexico. *Miscellaneous Publications of the Entomological Society of America*, **62**, 1–68.

Hall, M.J.R. and Wall, R. (1995) Myiasis of humans and domestic animals. *Advances in Parasitology*, **35**, 257–334.

Kersten, R., Shoukrey, N.M. and Tabbara, K.F. (1986) Orbital myiasis. *Ophthalmology*, **93**, 1228–32.

Krafsur, E.S., Whitten, C.J. and Novy, J.E. (1987) Screw-worm eradication in North and Central America. *Parasitology Today*, **3**, 131–7.

Lane, R.P. (1987) Flies causing myiasis. In *Manson's Tropical Diseases*, ed. P.E.C. Manson-Bahr and D.R. Bell, pp. 1462–8. London: Baillière Tindall.

Lane, R.P., Lovell, C.R., Griffiths, W.A.D. and Sonnex, T.S. (1987) Human cutaneous myiasis – a review and report of three cases due to *Dermatobia hominis*. *Clinical and Experimental Dermatology*, **12**, 40–5.

Nunzi, E., Rongioletti, F. and Rebora, A. (1986) Removal of *Dermatobia hominis* larvae. *Archives of Dermatology*, **122**, 140.

Sherman, R.A., Hall, M.J.R. and Thomas, S. (2000) Medicinal maggots: an ancient remedy for some modern afflictions. *Annual Review of Entomology*, **45** (in press).

Sherman, R.A. and Pechter, E.A. (1988) Maggot therapy: a review of the therapeutic application of fly larvae in human medicine, especially for treating osteomyelitis. *Medical and Veterinary Entomology*, **2**, 225–30.

Smith, K.G.V. (1986) *A Manual of Forensic Entomology*, pp. 93–137. London: British Museum (Natural History) and Ithaca, New York: Cornell University Press.

Snow, J.W., Siebenaler, A.J. and Newell, F.G. (1981) Annotated bibliography of the screw-worm, *Cochliomyia hominivorax* (Coquerel). *United States Department of Agriculture, Science and Education Administration, Agricultural Reviews and Manuals (Southern Series)*, **14**, 1–32.

Spradbery, J.P. (1991) *A Manual for the Diagnosis of Screw-worm Fly*. Canberra: Commonwealth Scientific and Industrial Research Organization, Division of Entomology.

Zumpt, F. (1965) *Myiasis in Man and Animals in the Old World: A Textbook for Physicians, Veterinarians and Zoologists*. London: Butterworth.

11

Fleas (Siphonaptera)

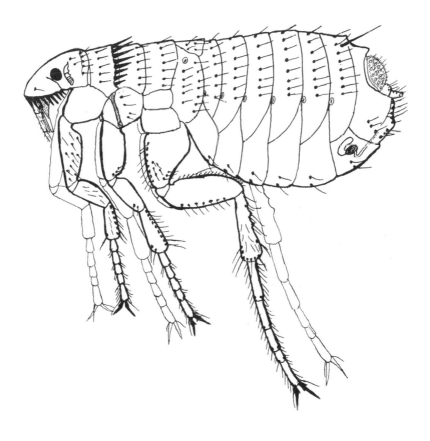

There are some 2500 species and subspecies of fleas belonging to 239 genera, but only relatively few are important pests of humans. About 94% of known species bite mammals and the remainder are parasitic on birds. Fleas are found throughout most of the world, but many genera and species have a more restricted distribution; for example the genus *Xenopsylla*, which contains important vectors of plague, is confined to the tropics and warmer parts of some temperate countries.

Medically the most important fleas are *Xenopsylla* species, such as *X. cheopis*, which is a vector of plague (*Yersinia pestis*) and flea-borne endemic typhus (*Rickettsia typhi*). Some fleas, such as *Ctenocephalides* species, are intermediate hosts of cestodes (*Dipylidium caninum*, *Hymenolepis diminuta*). Fleas may also be vectors of tularaemia (*Francisella tularensis*), and the chigoe or jigger flea (*Tunga penetrans*) burrows into the feet of people.

11.1   EXTERNAL MORPHOLOGY

Adult fleas are relatively small (1–4 mm), more or less oval insects, compressed laterally and varying in colour from light to dark brown. Wings are absent, but there are three pairs of powerful and well-developed legs, the hind pair of which are specialized for jumping. The legs, and also much of the body, are covered with bristles and small spines.

The head is roughly triangular in shape, bears a pair of conspicuous black eyes (a few species are eyeless), and short three-segmented more or less club-shaped antennae which lie in depressions behind the eyes. The mouthparts point downwards. In some species a row of coarse, well-developed and tooth-like spines, collectively known as the genal comb or genal ctenidium, is present along the bottom margin of the head capsule (Figs. 11.1, 11.2).

The thorax has three distinct segments: the pro-, meso- and metathorax. The posterior margin of the pronotum may bear a row of tooth-like coarse spines forming the pronotal comb or pronotal ctenidium (Fig. 11.1). Some genera of fleas lack both the pronotal and genal combs and are referred to as combless fleas (Fig. 11.2), whereas in other genera both combs are present (Fig. 11.2). In some species the pronotal comb is present and the genal comb absent (Fig. 11.2), but never the reverse. A sternite called the mesopleuron (mesosternum) is located above the middle pair of legs. In several genera, including *Xenopsylla*, which contains important plague vectors, this sternite is clearly divided into two parts by a thick vertical rod-like structure called the pleural rod, meral rod or mesopleural suture. The presence of this rod, combined with the absence of both genal and pronotal combs, indicates the genus *Xenopsylla* (Fig. 11.2). However, it must be stressed that the presence of a meral rod by itself does not identify fleas as species of *Xenopsylla*, because several other genera which have combs also have a meral rod.

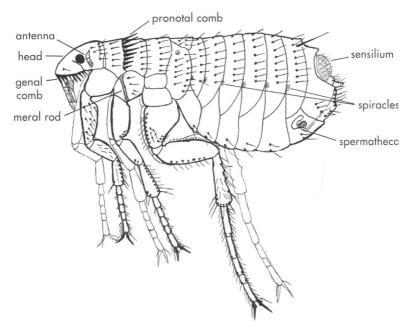

Figure 11.1 Lateral view of an adult flea showing the positions of combs and meral rod (pleural rod).

Male fleas are identified by the upturned appearance of the abdomen. In the female the tip of the abdomen is more rounded than in the male, and lying internally in about the position of the sixth to eighth abdominal segments are one or two distinct brownish spermathecae (Fig. 11.1). It is not always important to separate the sexes because both take blood-meals and can be vectors of disease. A flat or convex plate, called the sensilium or pygidium, and having spicules and dome-shaped structures, is present dorsally on segment 8 of both sexes. This structure is characteristic of fleas, but its exact function is unknown, although it is clearly sensory.

11.2   THE ALIMENTARY CANAL OF ADULT FLEAS

For a better understanding of the role fleas have in the transmission of plague it is necessary to describe the alimentary canal and the method of blood-feeding.

During feeding saliva is injected and the host's blood is sucked up through the spindle-shaped pharynx and thin oesophagus into the bulbous proventriculus (Fig. 11.3). This is provided internally with numerous (250–450) backwardly projecting stiff spine-like setae which when pressed together prevent the regurgitation of the blood-meal into the oesophagus (Fig. 11.3). (The proventriculus is important in the mechanism of plague transmission.) Finally, the blood-meal enters a relatively large

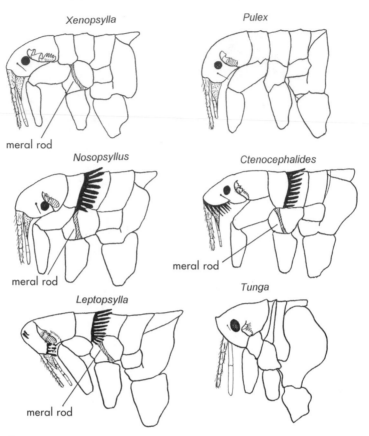

Figure 11.2 Diagram of the head and first three thoracic segments of adult fleas showing features that distinguish six medically important genera.

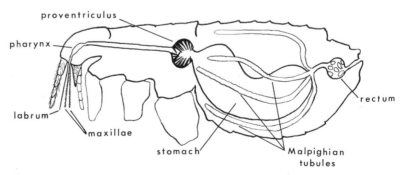

Figure 11.3 Diagrammatic representation of the alimentary canal of an adult flea showing backwardly projecting spines in the proventriculus.

stomach (mid-gut) where it is digested. The distal end of the stomach is connected to the hind-gut, the junction being easily identified by the four Malpighian tubules. The hind-gut is continuous with a small dilated rectum which has prominent rectal papillae. These papillae extract water from the faeces, so that they pass out through the anus in an almost dry state.

11.3   LIFE CYCLE

Both sexes of fleas take blood-meals and are therefore equally important as vectors of disease. The present account is a generalized description of the life cycle of fleas which may occur on humans or animals, such as dogs, cats and commensal rats. The life cycle of the chigoe (*Tunga penetrans*) is described separately.

A female flea which is ready to oviposit may leave the host to deposit her eggs in debris which accumulates in the host's dwelling place, such as rodent burrows or nests. With species which occur on humans or their domestic pets, such as cats and dogs, females often lay their eggs in or near cracks and crevices on the floor or amongst dust, dirt and debris. Sometimes, however, eggs are laid while the flea is still on the host and these usually, but not always, fall to the ground. The eggs are very small, oval, white or yellowish and lack any pattern. They are thinly coated with a sticky substance which usually results in them becoming covered with dirt and debris. Adults commonly live for 10 days to 6 weeks, but sometimes for 6–12 months or even longer. During this time a female may lay 300–1000 eggs, mostly in small batches of 3–25 a day.

Eggs hatch within 2–14 days, mostly after 5 days, but this depends on the species of flea, temperature and humidity. A minute legless larva emerges from the egg (Fig. 11.4). It has a small blackish head with a pair of very small antennae, followed by 13 pale-brown, distinct and more or less similar segments. Each segment bears a circle of setae near the posterior border. The last segment ends in a pair of finger-like ventral processes termed the anal struts. The presence of these struts, combined with the setae on the body, distinguish larval fleas from all other types of insects of medical importance.

Larvae are very active. They avoid light and seek shelter in cracks and crevices and amongst debris on floors of houses, or at the bottom of nests and animal burrows. Sometimes, however, larvae are found amongst the fur of animals, and even on people who have unclean habits and dirt-laden clothes, and occasionally in beds. Larvae feed on almost any organic debris including the host's faeces, and partly digested blood evacuated from the alimentary canal of adult fleas; in a few species feeding on expelled blood seems to be a nutritional requirement for larval development. In some

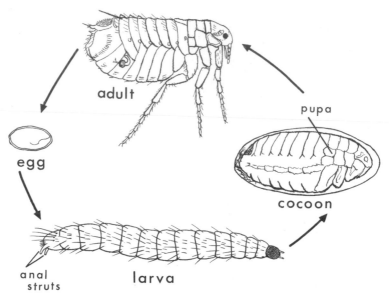

Figure 11.4 Life cycle of a flea.

species larvae are scavengers and feed on small dead insects or dead adult fleas of their own kind. There are usually three larval instars, but in a few species there are only two instars (e.g. *Tunga penetrans*, p. 177). The larval period may last as little as 10–21 days, but varies greatly according to species, and may be prolonged to more than 200 days by unfavourable conditions such as limited food supply and low temperatures. Mature larvae are 4–10 mm long. Unlike adult fleas, larvae cannot tolerate large extremes in relative humidity and they die if humidities are either too low or too high.

At the end of the larval period the larva spins a whitish cocoon from silk produced by the larval salivary glands. Because of its sticky nature it soon becomes covered with fine particles of dust, organic debris and sand picked up from the floor of the host's home. Cocoons camouflaged in this way are very difficult to distinguish from their surroundings. About 2–3 days after having spun a cocoon around itself the larva pupates within the cocoon. Adults emerge from the pupa after about 5–14 days, but this period depends on the ambient temperature. After having emerged from the pupa the adult flea requires a stimulus, usually vibrations, such as those caused by the movements of the host within its home, burrow or nest, before it escapes from the cocoon. If, however, animal shelters or houses are vacated, then adult fleas fail to escape from their cocoons until their dwelling places are reoccupied. In some species, carbon dioxide emitted from hosts or a seasonal increase in humidity stimulates emergence. Adults may remain alive in their cocoons for about 1 year. This explains why people moving into buildings which have been vacated for many months may suddenly be attacked by large numbers of very bloodthirsty fleas, these being newly emerged adults seeking their first blood-meal.

The life cycle from egg to adult emergence may be as short as 2–3 weeks for certain species under optimum conditions, but frequently the life cycle is considerably longer, taking many months.

Fleas avoid light and are therefore usually found sheltering amongst the hairs or feathers of animals, or on people under their clothing or in the bed. Given the opportunity many species of fleas feed several times during the day or night on their hosts. While feeding, fleas eject faeces composed at first of semidigested blood of the previous meal and then excess blood taken in during the act of feeding. This mixture of partially digested and virtually undigested blood often marks clothing and bed linen of people heavily infected with fleas.

Although most species of fleas have one or two favourite species of hosts, they are not entirely host-specific; for example, cat and dog fleas (*Ctenocephalides felis* and *C. canis*) will readily feed on humans, especially in the absence of their normal hosts. Human fleas (*Pulex irritans*) feed on pigs, and rat fleas of the genus *Xenopsylla* will attack people in the absence of rats. Most fleas will in fact bite other hosts in their immediate vicinity when their normal hosts are absent or scarce. However, although feeding on less acceptable hosts keeps fleas alive, their fertility can be seriously reduced by continued feeding on such hosts. Fleas rapidly abandon dead hosts to seek out new ones, behaviour which is of profound epidemiological importance in plague transmission. Fleas can withstand both considerable desiccation and prolonged periods of starvation, for example 6 months or more, when no suitable hosts are present. On their host, fleas move either by rapidly crawling or by jumping, but off the host they tend to jump more than crawl in their search for new hosts. Fleas can jump about 20 cm vertically and 30 cm or more horizontally. (Such remarkable feats are achieved through a rubber-like protein called resilin, which is very elastic and can become highly compressed; rapid expansion of the compressed state gives the power for jumping.)

11.4 MEDICAL IMPORTANCE

11.4.1 Flea nuisance

Although certain species of fleas may be important vectors of disease, the most widespread complaint about them concerns the annoyance caused by their bites, which in some people lead to considerable discomfort and irritation. The most common nuisance flea is the cat flea, *Ctenocephalides felis* (four subspecies of *C. felis* are recognized; the true cat flea is *C. felis felis*, but for convenience will be referred to as just *C. felis*). Of lesser importance as a pest is the dog flea, *Ctenocephalides canis*, and morely rarely the so-called human flea, *Pulex irritans*. In some areas other species such as the European chicken flea (*Ceratophyllus gallinae*) and the Western chicken flea (*Ceratophyllus niger*) may be of local importance.

Fleas frequently bite people on the ankles and legs, but at night a sleeping person is bitten on other parts of the body. In many people the bite is felt almost immediately, but irritation usually becomes worse some time after biting. In sensitized people intense itching may result. Because fleas are difficult to catch this tends to increase the annoyance they cause, and people attacked by fleas frequently spend sleepless nights alternately scratching themselves and trying to catch the fleas. There is evidence that children under 10 years generally experience greater discomfort than older people.

11.4.2 Plague

Plague is caused by *Yersinia pestis* and is primarily a disease of wild animals, especially rodents, not people. Over 220 rodent species have been shown to harbour plague bacilli. The cycle of transmission of plague between wild rodents, such as gerbils, marmots, voles, chipmunks and ground squirrels, is termed sylvatic, campestral, rural or endemic plague. Many different species of fleas bite these rodents and maintain plague transmission amongst them. When people such as fur trappers and hunters handle these wild animals there is the risk that they will get bitten by rodent fleas and become infected with plague.

An important form of plague is urban plague. This describes the situation when plague circulating among the wild rodent population has been transmitted to commensal rats, and is maintained in the rat population by fleas such as *Xenopsylla cheopis* (Europe, Asia, Africa and the Americas), *X. astia* (South-east Asia) and *X. brasiliensis* (Africa, South America and India). When rats are living in close association with people, such as in rat-infested slums, fleas normally feeding on rats may turn their attention to humans. This is most likely to happen when rats are infected with plague and as a result rapidly develop an acute and fatal septicaemia. On death of the rats the infected fleas leave their more normal hosts and feed on humans. In this way bubonic plague is spread by rat fleas to the human population. The most important vector species is *X. cheopis* but other species, including *Xenopsylla astia* and *X. brasiliensis*, and more rarely *Nosopsyllus fasciatus* and *Leptopsylla aethiopica*, have been incriminated in plague transmission in some areas. However, the last two rodent fleas are reluctant to feed on people and so rarely transmit the disease to humans. In addition to humans becoming infected by the bite of fleas which have previously fed on infected rats, the disease can also be spread from person to person by fleas, such as *Xenopsylla* species and *Pulex irritans*, feeding on a plague victim then on another person. This latter method, however, appears to play a minor role in the transmission of plague.

It is important to understand the methods by which fleas transmit plague. Plague bacilli sucked up with the blood-meals of male and female fleas are passed to the stomach where they undergo so great a multiplication that they extend forwards to invade the proventriculus. In some species, especially those of the genus *Xenopsylla*, further multiplication in the proventriculus may result in it becoming partially, or more or less completely, blocked. This prevents the proventriculus from functioning normally and results in fleas regurgitating some of the blood-meal during later feeds. Thus, plague bacilli obtained from a previous feed are passed down the flea's mouthparts into the host. In the case of completely blocked fleas, blood is sucked up with considerable difficulty about as far as the proventriculus, where it mixes with the bacilli and is then regurgitated back into the new host. Blocked fleas soon become starved and repeatedly bite in attempts to get a blood-meal, and are therefore potentially very dangerous. Even if there is no blockage of the alimentary canal, plague transmission can nevertheless occur by direct contamination from the flea's mouthparts.

Another, but less important, method of infection is by the flea's faeces being rubbed into abrasions in the skin or coming into contact with mucous membranes. Plague bacilli can remain infective in flea faeces for as long as 3 years. Occasionally the tonsils become infected with plague bacilli due to crushing infected fleas between the teeth.

In pneumonic plague the bacilli occur in enormous numbers in the sputum. This form of plague is highly contagious due to the ease by which bacilli are transmitted from patients to others by coughing, and the inhalation of droplets; insects are not involved in the spread of pneumonic plague.

11.4.3 Flea-borne endemic typhus

Flea-borne or murine typhus is caused by *Rickettsia typhi* (=*mooseri*) which is ingested by the flea with its blood-meal. Within the gut the rickettsiae multiply, but unlike plague bacilli they do not cause any blockage of the proventriculus or stomach. Infection is caused by infected faeces being rubbed into abrasions or coming into contact with delicate mucous membranes, and also by the release of rickettsiae from crushed fleas. Faeces may remain infective for 1–3 months. Murine typhus is essentially a disease of rodents, particularly rats and especially *Rattus rattus* and *R. norvegicus*. It is spread among rats and other rodents by *Xenopsylla* species, especially *X. cheopis*, but also by *Nosopsyllus fasciatus*, *Leptopsylla segnis* and by a few ectoparasites which are not fleas, such as the rat louse *Polyplax spinulosa* (possibly also *Hoplopleura* species) and maybe by the tropical rat mite *Ornithonyssus bacoti*. People become infected mainly through *Xenopsylla cheopis*, but occasionally *Nosopsyllus fasciatus*, *Ctenocephalides canis*, *C. felis* and *Pulex irritans* may be involved. *Leptopsylla segnis* does not attack humans, but it is possible that murine typhus is sometimes spread to people by an aerosol of infective faeces of this flea.

The rickettsiae of murine typhus can pass across to the flea's ovaries, into the eggs and then to the larvae, in a form of trans-ovarial transmission. But whether this is epidemiologically important remains unknown.

11.4.4 Cestodes

Dipylidium caninum is the commonest tapeworm of dogs and cats and occasionally occurs in children. It can be transmitted by fleas to both rodents and humans in the following way. Eggs of the tapeworm are passed out with the excreta of domestic pets and may be swallowed by larval fleas feeding on the excreta. Larval worms hatching from the ingested eggs penetrate the gut wall of the larval flea and pass across into the body cavity (coelom). They remain trapped within this space and pass on to the pupa and finally to the adult flea where they encapsulate and become cysticercoids (infective larvae). Animals can become infected by licking their coats during grooming and thus swallowing the infected adult fleas. Similarly, young children fondling and kissing dogs and cats can become infected with *D. caninum* by swallowing cat and dog fleas, or by being licked by dogs which have crushed infected fleas in their mouths thus liberating the infective cysticercoids.

The rat tapeworm, *Hymenolepis diminuta*, of rats and mice has a similar life cycle. Very occasionally adult rat fleas get mixed with food and drink and are swallowed by humans, who may then become infected with the tapeworm.

11.4.5 Less important diseases

In addition to the above diseases and parasites spread by fleas these insects may play some small part in the transmission of *Francisella tularensis*, *Rickettsia conori*, *Coxiella burneti*, *Bartonella henselae* (cat scratch disease) and a few other rather minor infections. However, it must be stressed that the role of fleas as vectors of these pathogens is minimal.

11.5   *TUNGA PENETRANS*

11.5.1 External morphology

Tunga penetrans is found in the tropics and subtropics, having a distribution stretching from Central and South America, the West Indies across Africa

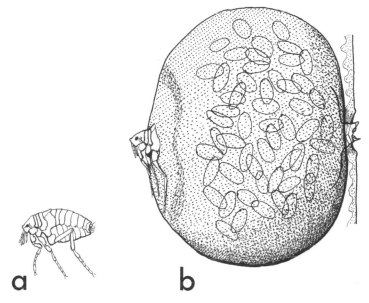

Figure 11.5 Adults of *Tunga penetrans*. (*a*) Non-gravid female flea; (*b*) gravid female with enormously swollen abdomen full of eggs, embedded in the skin of a host. Note (on right) that the tip of the abdomen projects from the host's skin to the exterior.

to Madagascar. It has occasionally been reported in persons in India returning from overseas, mainly Africa, but the flea is not indigenous to India or elsewhere in Asia. *Tunga penetrans* is sometimes referred to as the chigoe or jigger or sand-flea. *Tunga penetrans* does not transmit any disease to people but is a nuisance because females burrow into the skin.

Adults of both sexes are exceedingly small, only about 1 mm long (Fig. 11.5*a*). They have neither genal nor pronotal combs and are easily separated from other fleas of medical importance by their very compressed first three (thoracic) segments, and the paucity of spines and bristles on the body.

11.5.2 Life cycle and medical importance

Eggs of *Tunga penetrans* are dropped onto the floor of houses or on the ground outside. They hatch within about 3–4 days and the larvae inhabit dirty and dusty floors or dry sandy soils, especially in areas frequented by hosts of the adult fleas. There are two instars, not three as in most flea species, and under favourable conditions larval development is completed within about 10–14 days; the pupal period lasts about 5–14 days. The complete life cycle can be as short as about 18 days.

Newly emerged adults are very agile and jump and crawl about on the ground until they locate a suitable host, which is usually a person or pig. Both sexes feed on blood, but whereas the male soon leaves the host after taking a blood-meal, the female, after being fertilized, burrows into the skin where it is soft, such as between the toes or under toe-nails. Other areas of the foot, including the sole, may also be invaded. In people habitually sitting on the ground, such as beggars or infants, the buttocks may often be infected, and particularly heavy infestations have been recorded from leprosy patients. In heavily infected individuals the arms, especially the elbows, may also be attacked, and occasionally the females burrow into the soft skin around the genital region. Burrowing into the skin appears to be accomplished by the flea's sharp and well-developed mouthparts. The result is that the entire flea, with the exception of the tip of the abdomen bearing the anus, genital opening and large respiratory spiracles, becomes completely buried in the host's skin. In this embedded position she continues to feed. The area surrounding the embedded flea becomes very itchy and inflamed, and secondary infections may become established, resulting in ulcerations and accumulation of pus. While the blood-meal is being digested, the abdomen distends to an enormous size (about 1000 times its original size) and the flea attains both the shape and size (6 mm) of a small pea (Fig. 11.5b). This expansion is accomplished in some 8–10 days. Towards the end of this period of abdominal enlargement the ovaries are composed of thousands of minute eggs. Over the next 7–14 days about 150–200 eggs are passed out of the female genital opening, most of which eventually fall to the ground and hatch after about 3–4 days.

When the female fleas die they remain embedded within the host. This frequently causes inflammation and may, in addition, result in secondary infections, which if ignored can lead to loss of the toes, tetanus, or even gangrene. Male fleas cause no such trouble as they do not burrow into the skin.

These fleas are most common in people not wearing shoes, such as children. Because the fleas are feeble jumpers, wearing shoes is a simple, but in some communities relatively costly, method of reducing the likelihood of flea infection.

Females embedded in the skin should be removed with fine needles under aseptic conditions, and wounds caused by their extraction sterilized and dressed. They are best removed within the first few days of their becoming established, as when they have greatly distended abdomens, containing numerous eggs, they are difficult to extract without rupturing them, and this increases the risk of infections.

Pigs, in addition to humans, are often commonly invaded by *Tunga*, and may provide a local reservoir of infection. Other animals such as cats, dogs and rats are also readily attacked.

11.6   CONTROL OF FLEAS

Repellents such as dimethylphthalate (dimp), diethyltoluamide (deet) or benzyl benzoate may afford some personal protection against fleas.

Resistance has been reported in cat fleas and *Xenopsylla* species in various parts of the world to one or more of the following categories of insecticides: organochlorines, organophosphates, carbamates, pyrethrins and pyrethroids. Nevertheless insecticides remain the main tool for flea control, although there is increasing reliance on the use of insect growth regulators (IGRs).

11.6.1   Cat and dog fleas

Cat and dog fleas (*Ctenocephalides felis* and *C. canis*) can most easily be detected by examination of the fur round the neck, or on the belly of the hosts. Proprietary insecticidal powders, sprays or shampoos containing 1% HCH, 2–5% malathion, 2–5% carbaryl (Sevin), 1–2% diazinon, 0.5–1% permethrin, or other insecticides can be applied to the animal's fur. Dusts are safer to use than liquids because they are less likely to be absorbed through the animal's skin and cause unpleasant side-reactions. A simple, but not always very effective, procedure is to place a plastic collar impregnated with 20% dichlorvos (DDVP), 10% propoxur (Baygon), carbaryl (Sevin), fenthion (Baytex) or temephos (Abate), or an IGR such as methroprene, around the necks of cats or dogs. Flea collars remain insecticidal for 2–3 months, or up to a year with methoprene, but they may cause skin problems in pets. Alternatively insecticides such as permethrin or fenthion can be formulated as a 'spot-on' solution that is applied to the pet's skin in just one small area. The insecticide is absorbed through the skin and passes to the animal's blood, so that blood-feeding fleas ingest the insecticide. One treatment lasts for about 3–4 weeks. IGRs such as methoprene, fenoxycarb or pyriproxyfen can also be applied topically. They provide prolonged ovicidal activity.

When IGRs such as cryomazine or lufenuron are administered orally once a month, fleas feeding on the animals produce non-viable eggs for 2–3 weeks, and effective flea control may last for about 4–6 weeks.

However, an important consideration is that most fleas are found away from the host, not on it. For example, it has been said that a typical colony of cat fleas consists of only about 25 adult fleas on the cat, but that on the floor and bedding there may be 500 adult fleas, 500 cocoons and as many as 3000 larvae and 1000 eggs. Clearly, control measures should not be restricted to the cat but applied to the total environment. Flea cocoons are not very susceptible to insecticides; consequently insecticidal treatments should be repeated about every 2 weeks for about 6 months. Beds, kennels, or other places where pets sleep, or spend much of their time, should be treated either with insecticidal powders or lightly sprayed with solutions containing 0.5% HCH, 1–2% dichlorvos (DDVP), 2–5% malathion, 0.5% diazinon, 0.5% chlorpyrifos (Dursban) or one of the pyrethoids, to kill both adults and larval fleas. More recently IGRs such as methoprene, fenoxycarb or pyriproxyfen have been sprayed onto carpets, floors and pets' bedding. Duration of effective control depends on the types of materials sprayed (e.g. earthen or wooden floors, synthetic or woollen carpets), but IGRs generally remain effective longer that conventional insecticides, some giving good control for up to 2 months.

Vacuum cleaning floors, carpets and pets' bedding can also be very effective in removing the immature stages of fleas.

11.6.2 Rodent fleas

For more general control of fleas, powders of 5–10% DDT, 1% HCH or 0.5% dieldrin can be liberally applied to floors of houses and runways of rodents. Insecticidal dusts can also be blown into rodent burrows. In many parts of the world, however, *Xenopsylla cheopis* and *Pulex irritans* have developed resistance to DDT, HCH and dieldrin. In such cases organophosphate or carbamate insecticides such as 2% diazinon, 2% fenthion (Baytex), 5% malathion, 2% fenitrothion (Sumithion), 5% iodofenphos (jodfenphos) or 3–5% carbaryl (Sevin) can be used. In India *X. cheopis* has developed resistance to malathion, and trials in that country have shown carbaryl (Sevin) to be the most effective insecticide.

Insecticidal fogs or aerosols containing 2% malathion or 2% fenchlorphos (Ronnel) have sometimes been used to fumigate houses harbouring fleas. Insecticidal smoke bombs containing HCH, DDT, permethrin or pirimiphos methyl (Actellic) can also be used to disinfest houses.

When houses or outdoor areas are sprayed with IGRs, such as pyriproxyfen or fenoxycarb, longer-lasting control (for at least 4 months) of cat fleas is achieved than when insecticides are used. Although IGRs appear to have good potential for control of plague fleas they have received little evaluation.

For the control of fleas in urban outbreaks of plague or murine typhus, extensive and well-organized insecticidal operations may be necessary. At the same time as insecticides are applied, rodenticides such as the anticoagulants warfarin and fumarin can be administered to kill the rodent population. However, if fast-acting 'one-dose' rodenticides such as zinc phosphide, sodium fluoroacetate or strychnine or the more modern fast-acting anticoagulants such as bromadiolone and chlorophacinone are used, then it is essential to apply these several days after insecticidal applications. Otherwise the rodents will be killed but not their fleas, which will then bite other mammals including people, and this may result in increasing disease transmission.

FURTHER READING

Azad, A.F. (1990) Epidemiology of murine typhus. *Annual Review of Entomology*, **35**, 553–69.

Gratz, N.G. (1980) Problems and developments in the control of flea vectors of disease. In *Fleas*, ed. R. Traub and H. Starcke, pp. 217–40. Proceedings of the International Conference on Fleas, Ashton Wold, Peterborough, UK, 21–25 June 1977. Rotterdam: Balkema.

Gratz, N.G. and Brown, A.W.A. (1983) XII. Fleas: biology and control. WHO/VBC/83.874. Geneva: World Health Organization mimeographed document.

Hinkle, N.C., Koehler, P.G. and Patterson, R.S. (1995) Residual effectiveness of insect growth regulators applied to carpet for control of cat flea (Siphonaptera: Pulicidae) larvae. *Journal of Economic Entomology*, **88**, 903–6.

Hirst, L.F. (1953) *The Conquest of Plague: A Study of the Evolution of Epidemiology*. Oxford: Clarendon Press.

Jellison, W.L. (1959) Fleas and disease. *Annual Review of Entomology*, **4**, 389–414.

Pugh, R.E. (1987) Effects on the development of *Dipylidium caninum* and on the host reaction to this parasite in the adult flea (*Ctenocephalides felis felis*). *Parasitological Research*, **73**, 171–7.

Rothschild, M. (1975) Recent advances in our knowledge of the order Siphonaptera. *Annual Review of Entomology*, **20**, 241–59.

Rust, M.K. and Dryden, M.W. (1997) The biology, ecology, and management of the cat flea. *Annual Review of Entomology*, **42**, 451–73.

Schriefer, M.E., Sacci, J.B., Taylor, J.P., Higgins, J.A. and Azad, A.F. (1994). Murine typhus: updated roles of multiple urban components and a second typhuslike rickettsia. *Journal of Medical Entomology*, **31**, 681–5.

Traub, R. and Starcke, H. (eds.) (1980) *Fleas*. Proceedings of the International Conference on Fleas, Ashton Wold, Peterborough, UK, 21–25 June 1977. Rotterdam: Balkema.

Traub, R., Wisseman, C.L. and Farhang-Azad, A. (1978) The ecology of murine typhus: a critical review. *Tropical Diseases Bulletin*, **75**, 237–317.

12

Lice (Anoplura)

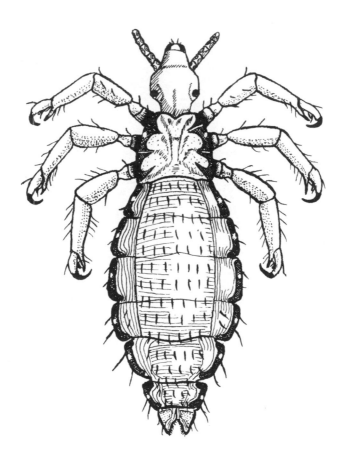

Blood-sucking lice on humans comprise three types, namely the pubic or crab louse (*Pthirus pubis*), the body louse (*Pediculus humanus*) and the head louse (*Pediculus capitis*). Morphologically it is very difficult to separate the head and body louse. The two can interbreed, and the head louse has sometimes been regarded as a subspecies of the body louse, but it is now usually treated as a separate species. All three species of lice have a more or less world-wide distribution, but they are often more common in temperate areas.

These three species of lice belong to the Anoplura, which has variously been regarded as a separate order, or a suborder of the order Mallophaga, which contains the chewing lice, insects commonly found on domestic pets and other animals. Here Anoplura is treated as a distinct order.

Body lice are vectors of louse-borne typhus (*Rickettsia prowazeki*), trench fever (*Bartonella quintana*; previously in the genera *Rochalimaea* and *Ricksettsia*) and louse-borne relapsing fever (*Borrelia recurrentis*).

12.1   THE BODY LOUSE (*PEDICULUS HUMANUS*)

12.1.1 External morphology

Adults are small, pale beige or greyish wingless insects, with a soft but rather leathery integument, and are flattened dorsoventrally (Fig. 12.1). Males measure about 2–3 mm and females about 3–4 mm. The head bears a pair of inconspicuous eyes and a pair of short five-segmented antennae.

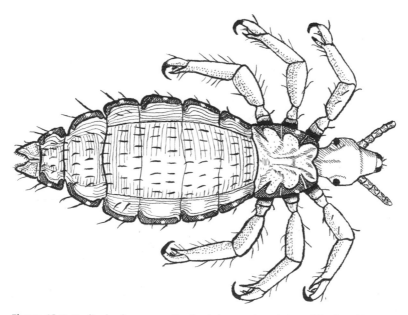

Figure 12.1 *Pediculus humanus*, the body louse, dorsal view. (The head louse, *Pediculus capitis*, looks virtually identical.)

The three pairs of legs are stout and well developed. The short thick tibia has a small thumb-like spine on its inner side at the apex, and the short tarsus has a curved claw (Fig. 12.2a). In males the claw on the fore-legs is longer than in the females. Hairs of the host, or clothing, are gripped between this spine and claw.

The mouthparts of the louse are different from those of most other blood-sucking insects in that they do not constitute a projecting piercing proboscis. They consist of a flexible, sucking, almost tube-like mouth, called the haustellum, which is armed on the inner surface with minute teeth which grip the host's skin during feeding. Needle-like stylets are thrust into the skin and saliva is injected into the wound. Blood is sucked into the mouth and passes into the stomach for digestion.

The lateral margins of the abdominal segments are sclerotized and much darker than the rest of the segments.

In the males there are dark transverse bands on the dorsal surface of the abdominal segments and the tip of the abdomen is rounded, whereas in females it is bifurcated. This is associated with the female gripping fibres of clothing during egg laying.

12.1.2 Life cycle

Both sexes take blood-meals and feeding occurs at any time during the day or night. Both the adult and immature stages live permanently on humans, clinging mainly to hairs of their clothing and usually only to body hairs during feeding. Female lice glue about 6–9 eggs per day very firmly on to the hairs of clothing, especially to those along the seams of underclothes, such as vests and pants, but also on shirts and occasionally on body hairs. The egg, commonly called a nit (though strictly this term should be applied only to the hatched egg), is oval, white, about 1 mm long, and has a distinct operculum (cap) containing numerous small perforations which give the egg the appearance of a miniscule pepper pot (Fig. 12.2b, c). The intake of air through these holes not only supplies the tissues of the developing embryo with oxygen but aids hatching in the following way. Just prior to hatching, the fully developed louse within its eggshell swallows air which distends the body against the eggshell, thus building up a back pressure causing the head of the louse to be pushed up against the operculum and forcing it off. Female lice live for 2–4 weeks and lay 200–300 eggs.

The duration of the egg stage is normally about 7–10 days, but eggs on discarded clothing away from the warmth of the body may not hatch until 2–3 weeks; or in cool places not at all. Eggs cannot survive longer than 4 weeks; consequently there is little danger of infestation with body lice from clothing which has not been worn for over 1 month.

Lice have a hemimetabolous life cycle. The louse which hatches from the egg is termed a nymph and resembles a small adult louse. It takes a blood-

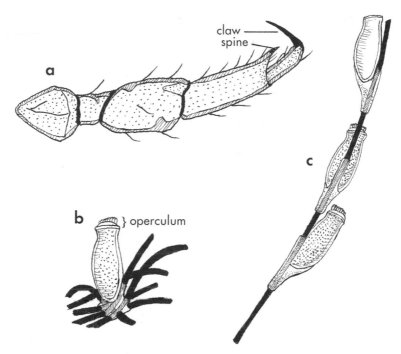

Figure 12.2 Head and body lice. (*a*) Leg of a body louse showing tarsal claw and tibial spine; (*b*) unhatched egg of a body louse glued to fibres of clothing; (*c*) two unhatched eggs (lower) and one hatched egg (upper) of the head louse cemented to a hair. For convenience these eggs are shown very close together, but in practice they are very rarely this close.

meal and passes through three nymphal instars, and after about 7–12 days becomes an adult male or female louse. The duration of the nymphal stages depends much on whether or not clothing is worn all the time. If it is discarded at night this may result in subjecting the nymphs to lower temperatures, thus slowing down their development. Each louse usually takes several blood-meals a day from its host.

The body louse is a true ectoparasite of humans. Unfed lice die within about 2–4 days if kept away from humans and without a blood-meal, but blood-fed individuals may survive for 5–10 days. Lice are very sensitive to changes in temperature. They quickly abandon a dead person, owing to cooling down of the body, to seek out new hosts; they also leave a person with a high temperature. They are unable to feed at temperatures above about 40 °C.

A very heavily infested person may have 400–500 lice on the clothing and body. There is one record of an estimated 10000 lice and 10000 eggs from a single shirt! Usually, however, fewer than 100 lice are found on any one individual, and many have considerably less than this.

Body lice are spread by close contact and are especially prevalent under conditions of overcrowding and in situations where people rarely wash or change their clothes. They are therefore commonly found on people in primitive jails, refugee camps and in trenches during wars, and also after disasters such as floods or earthquakes when people are forced to live in very overcrowded, and usually insanitary, conditions. People living in mountainous areas such as those in East Africa, Ethiopia, Sudan, Burundi, Nepal, India and Andean regions of South America, where cold weather necessitates wearing several layers of clothes which are rarely changed or washed, often have lice. In more developed countries body lice are found mainly on homeless people, and infestations may reach a peak in cold weather when several layers of woollen underclothes are worn.

12.1.3 Medical importance

Pediculosis
The presence of body, head or pubic lice on a person is sometimes referred to as pediculosis. The skin of people who habitually harbour large numbers of body lice may become pigmented and tough, a condition known as vagabond's disease or sometimes as morbus errorum.

Because lice feed several times a day, saliva is repeatedly injected into people harbouring lice, and toxic effects may lead to weariness, irritability or a pessimistic mood: the person feels lousy. Allergies such as severe itching may be caused by repeated inoculation of saliva, and if inhaled the faeces may produce symptoms reminiscent of hay fever.

Louse-borne epidemic typhus
The rickettsiae of louse-borne typhus, *Rickettsia prowazeki*, are ingested with the blood-meal taken by both male and female lice, and also their nymphs. They invade the epithelial cells lining the stomach of the louse where they multiply enormously and cause the cells to become greatly distended. About 4 days after the blood-meal the gut cells rupture and release the rickettsiae back into the lumen of the insect's intestine. Due to these injuries to the intestinal wall, the blood-meal may seep into the haemocoel of the louse, giving the body an overall reddish colour. The rickettsiae are passed out with the faeces of the louse, and people become infected when these are rubbed or scratched into abrasions, or come into contact with delicate mucous membranes such as the conjunctiva. Infection can also be caused by inhalation of the very fine powdered dry faeces. Alternatively, if a louse is crushed, such as by persistent scratching because of the irritation caused by its bites, the rickettsiae in the gut are released and may cause infection through abrasions, etc. The rickettsiae may remain alive and infective in dried lice faeces for about 70 days.

Humans, therefore, become infected with typhus either by the faeces of the louse or by crushing it, not by its bite. An unusual feature of louse-borne epidemic typhus is that it is a disease of the louse as well as of humans. The rupturing of the epithelial cells of the intestine, caused by the multiplication of the rickettsiae, frequently kills the louse after about 8–12 days. This may explain why people suffering from typhus are sometimes found with no, or remarkable few, lice on their bodies or clothing.

People are usually considered to be the reservoir of the disease. Asymptomatic carriers remain infective to body lice for many years. Recrudescences as Brill–Zinsser's disease, many years after the primary attack, may occur in a person and lead to the spread of epidemic typhus.

Trench fever

Trench fever is a relatively uncommon and non-fatal disease which was first noticed during World War I (1914–18) among soldiers in the trenches, and then reappeared in eastern Europe during World War II (1939–45). More recently cases have been reported from the USA and Europe, mainly from the homeless.

Trench fever is caused by *Bartonella quintana*, the bacteria being ingested by the louse during feeding. The pathogens are attached to the walls of the gut cells where they multiply; they do not penetrate the cells as do typhus rickettsiae and consequently they are not injurious to the louse. After 5–10 days the faeces are infected. Like typhus, the disease is conveyed to humans either by crushing the louse, or by its faeces coming into contact with skin abrasions or mucous membranes. The pathogens persist for many months, possibly even a year, in dried louse faeces, and it is suspected that infection may commonly arise from inhalation of the dust-like faeces. The disease may be contracted by those who have no lice but are handling louse-infected clothing contaminated with faeces.

Louse-borne epidemic relapsing fever

The causative agent of louse-borne epidemic relapsing fever, *Borrelia recurrentis*, is ingested with the louse's blood-meal from a person suffering from epidemic relapsing fever. Within about 24 hours all spirochaetes have disappeared from the lumen of the gut. Many have been destroyed, but the survivors have succeeded in passing through the stomach wall to the haemocoel, where they multiply greatly, reaching enormous numbers by days 10–12. The only way in which humans can be infected with louse-borne relapsing fever is by the louse being crushed and the released spirochaetes entering the body through abrasions or mucous membranes. The habit of some people of crushing lice between the finger-nails, or the even less desirable habit of killing them by cracking them with the teeth, is clearly dangerous if the lice are infected with relapsing fever or rickettsiae.

The method of transmission of epidemic relapsing fever must make it very rare for more than one person to be infected by any one infected louse. Hence epidemics of louse-borne relapsing fever will rarely occur unless there are large louse populations.

12.1.4 Control

The most obvious way to eradicate body lice from a person is by changing and washing the clothing in water hotter than 60 °C, preferably followed by ironing. In epidemic situations, however, such measures may be impractical and immediate reinfestation may occur, hence insecticides are usually used for louse control. In some parts of the world body lice have developed resistance to DDT, malathion and some pyrethroids.

Ten per cent DDT dust mixed with an inert carrier (talc) can be blown by a plunger-type duster at the rate of 30–50 g per person, between the body and underclothes. If DDT-resistant lice are present then dusts of 1% malathion, 2% temephos (Abate), 1% propoxur (Baygon) or 0.5% permethrin can be used. Since insecticidal dusts come into close and prolonged contact with people, it is essential that they have very low mammalian toxicities.

Impregnation of clothing with a pyrethroid may provide long-lasting protection against lice infestations, and such treated clothing may remain effective after even six to eight washings.

12.2   THE HEAD LOUSE (*PEDICULUS CAPITIS*)

12.2.1 External morphology

There are only very minor morphological differences separating body and head lice. In practice, these differences are not very important because lice found on clothing or on the body are invariably body lice, whereas those on the head are nearly always head lice.

12.2.2 Life cycle

The life cycle of the head louse is very similar to that of the body louse except the eggs (nits) are not laid on clothes but are cemented to the hairs of the head, usually at their base (Fig. 12.2c), especially behind and above the ears, and at the back of the neck. Usually a single egg is laid on each hair. The distance between the scalp and the furthest egg glued to a hair is often regarded as providing an approximate estimate of the duration of infestation on the basis that a human hair grows at a rate of about 1 cm per month. However, it appears that eggs may also be laid on long hairs when they are near or touching the scalp, so that unhatched eggs can sometimes be some distance from the base of the hair. Only very occasionally are eggs laid on hairs elsewhere on the body. Most individuals harbour only 10–20 head lice, but in very severe infestations the hair may become matted with a mixture of nits, nymphs, adults and exudates from pustules resulting from bites of the lice. In such cases bacterial and fungal infections may become established and an unpleasant crust formed on parts of the head, underneath which are masses of head lice. Empty, hatched eggs remain firmly cemented to the hairs of the head. A female lays about 6–8 eggs per day, amounting to about 50–150 eggs during her life-span, which is a little over 2–4 weeks. Eggs hatch within 7–10 days and the duration of the nymphal stages is about 7–10 days. Away from people head lice die within 2–3 days.

As with body lice, dissemination of head lice is only by close contact, such as children playing together and their heads frequently touching, or when people are crowded together such as in prisons or refugee camps. Use of other people's hats, scarves, combs, hair brushes, etc. has not been proved to spread head lice. The occasional head lice found on such articles or on the backs of chairs are moribund individuals that will not be able to survive and infect someone else. Infestations are often, but not always, greater in women than men, and usually highest in children.

12.2.3 Medical importance

In many areas of the world head lice are a serious public health problem, and in many countries prevalence has been increasing. Often there are higher infestation rates in homes that are overcrowded and where there is poor hygiene. There is no evidence that head lice are natural vectors of the diseases transmitted by body lice; typhus epidemics are always associated with body lice. However, under experimental conditions head lice can transmit both rickettsiae and spirochaetes. Head lice can transmit impetigo, the bacteria being ingested with the blood-meal and passing out unharmed with the faeces.

12.2.4 Control

Regular washing with soap and warm water may reduce the numbers of nymphs and adults on the hair, but will have no effect on the eggs, which are firmly glued to the bases of the hairs. Lice and nits can be removed by a steel or plastic comb which has very closely set, fine teeth. Alternatively, the head can be shaved. Regular combing with an ordinary comb, although not removing the eggs, may reduce the number of nymphs and adults.

Resistance to DDT, malathion, permethrin, phenothrin and some other pyrethroids has been reported in some parts of the world.

Insecticidal formulations for louse control include dusts, emulsions and lotions. The choice of formulation and insecticide depends on the availability of proprietary brands, preference of patients, degree of residual action required and costs. Although dusts (e.g. 10% DDT and 1% malathion) are efficient, they are not acceptable to most people because they give the head a greyish appearance and reveal that the person has lice. Preparations such as shampoos which are applied and then washed off after a few minutes are not usually very effective and are not recommended.

In some countries carbaryl (Sevin) and HCH (lindane) are no longer recommended for control of head lice because there is some evidence that they may pose a health risk. Alternative insecticides are emulsions of 1% permethrin, 0.2% phenothrin, 0.5% malathion, 0.3–0.4% bioallethrin or 2% propoxur (Baygon). Although some commercial preparations proclaim that lotions need only remain on the head for 10 minutes or 2 hours, it is better to leave the insecticidal lotion on the head for about 12 hours, e.g. overnight, before washing it off. Insecticidal shampoos are ineffective, and may even encourage the development of resistant populations of lice.

Although some insecticides such as malathion and permethrin are reputed to be ovicidal, a second treatment after an interval of 7–10 days is recommended whatever insecticide is used, because it is difficult to kill all eggs with just a single application. None of the compounds will remove eggs cemented to the hairs, but these can be removed with a louse comb.

Head lice infestations tend to disappear from people who sleep under pyrethroid-impregnated bed nets used to protect them from mosquito bites (see p. 49).

Experiments using orally administered ivermectin suggest the drug kills head lice, but as yet ivermectin has not been registered for control of human lice.

12.3   THE PUBIC LOUSE (*PTHIRUS PUBIS*)

12.3.1 External morphology
The pubic louse is generally smaller (1.5–2 mm) than *Pediculus* and is easily distinguished from it. In the pubic louse there is less differentiation between the thorax and abdomen; together they are almost round so that the body is nearly as broad as long. Whereas all three pairs of legs are more or less equal in size in body and head lice, in the pubic louse the front pair, although as long as the other two pairs, is much more slender and has smaller claws. In marked contrast, the middle and hind-legs have massive claws (Fig. 12.3). The presence of a broad squat body, very large claws on the middle and hind-legs, together with the characteristically more sluggish movements, has resulted in the pubic louse being aptly called the crab louse.

12.3.2 Life cycle
The life cycle of *Pthirus pubis* is very similar to that of *Pediculus*. Eggs take about 6–8 days to hatch and the nymphal stages last about 10–17 days. Females lay a total of about 150–200 eggs, which are slightly smaller than

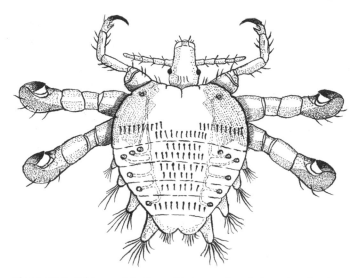

Figure 12.3 *Pthirus pubis*, the pubic louse, dorsal view, showing large tarsal claws of mid- and hind-legs.

those of the body and head louse and are cemented to the coarse hairs of the genital and perianal regions of the body. Unlike head lice several eggs may be laid on a single hair. Pubic lice may be found on other areas of the body having coarse and not very dense covering of hair, for example, the beard, moustache, eyelashes, underneath the arms and occasionally on the chest. They are very rarely found on the head. Pubic lice are considerably less active than *Pediculus*. The life cycle, from egg laying to formation of the adult, is about 17–25 days.

Infestation with crab lice is usually through sexual intercourse, and characteristically the French call them 'papillons d'amour'. However, it is wrong to suspect that this is the only method. Young children sleeping with parents can catch crab lice from them, and infestations can arise from discarded clothing, infested bedding, or even *rarely* from lavatory seats. Adults can survive only 2 days away from their hosts.

12.3.3 Medical importance
Although under laboratory conditions pubic lice can transmit louse-borne typhus, there is little evidence that under natural conditions they spread any disease, although there was a suggestion that they have been responsible for typhus outbreaks in China. In some individuals, however, severe allergic reactions (pruritus) develop to their bites due to the injection of saliva and the deposition around the feeding sites of faeces. Small characteristic bluish spots (maculae caeruleae) may appear on the infested parts

of the body. Infestations of pubic lice are sometimes known as pediculosis pubis or phthiriasis (note that, unlike the generic name of the louse, this is spelt with a 'phth').

12.3.4 Control

Originally control involved shaving pubic hairs from the body, but this method has been replaced by the application of insecticidal emulsions and lotions as used for head lice control.

Insecticide resistance has not been reported in pubic lice. Particularly effective is 1% permethrin, but a second treatment is advisable because eggs are not usually killed; however, a single application of an emulsion of 0.5% malathion should kill eggs as well as nymphs and adults.

It may be advisable to treat all hairy areas of the body below the neck. Insecticidal shampoos are ineffective for controlling pubic lice. Infestations on the eyelashes can be treated by applying a small amount of petroleum jelly along the closed lashes, twice a day for 10 days. This will kill the nymphs and adults; no attempt should be made to remove eggs as these will disappear as lashes fall out and are replaced by others. Alternatively the careful application of 0.75% physostigmine ophthalmic ointment can be used.

Some insecticides may cause irritation and dermatitis due to the sensitivity of the genital regions; in such cases shaving may remain a relatively simple and effective remedy.

FURTHER READING

Bradshaw, W. (1989) How prevalent are headlice? *Parasitology Today*, **5**, 135–6.

Burgess, I.F. (1995) Human lice and their management. *Advances in Parasitology*, **36**, 271–342.

Burgess, I.F. (1998) Head lice: developing a practical approach. *The Practitioner*, **242**, 126–9.

Buxton, P.A. (1948) *The Louse: An Account of the Lice which Infest Man, their Medical Importance and Control*, 2nd edn. London: Edward Arnold.

Chetwyn, K.N. (1996) An overview of mass disinfestation procedures as a means to prevent epidemic typhus. In *Proceedings of the 2nd International Conference on Insect Pests in the Urban Environment (ICIPUE)*, ed. K.B. Wildey, pp. 421–6. The Organising Committee of the ICIPUE.

Donnelly, E., Lipkin, J., Clore, E.R. and Altschuler, D.Z. (1991) Pediculosis prevention and control strategies of community health and school nurses. *Journal of Community Health Nursing*, **8**, 85–95.

Maunder, J.W. (1988) Updated community approach to head lice. *Journal of the Royal Society of Health*, **108**, 201–2.

Mumcuoglu, K.Y. (1996) Control of lice (Anoplura: Pediculidae) infestations: past and present. *American Entomologist*, **42**, 175–8.

Orkin, M. and Maibach, H.I. (eds.) (1985) *Cutaneous Infestations and Insect Bites*. See Chapters 19–26 on lice. New York: Marcel Dekker.

Pan American Health Organization (1973) Proceedings of the international
symposium on the control of lice and louse-borne diseases. *Pan American
Health Organization Scientific Publication*, **263**, 1–311.

Scholdt, L.L., Holloway, M.L. and Fronk, W.D. (1979) *The Epidemiology of Human
Pediculosis in Ethiopia*. Jacksonville, Florida: Navy Disease Vector Ecology and
Control Center.

Weidhaas, D.E. and Gratz, N.G. (1982) I. Lice. WHO/VBC/82.858. Geneva: World
Health Organization mimeographed document.

13

Bedbugs (Cimicidae)

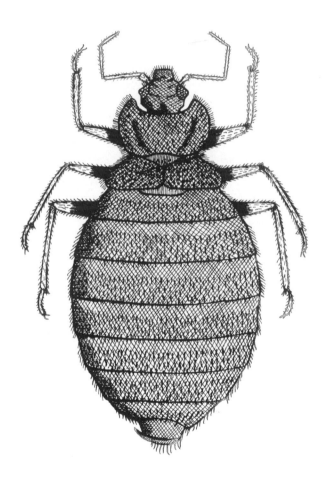

There are two common species of bedbugs, both of which feed on humans: *Cimex lectularius*, which has a widespread distribution in tropical and non-tropical countries and *C. hemipterus*, commonly called the tropical bedbug, which is essentially a species of the Old and New World tropics although it can also occur in hot areas of some non-tropical countries, such as Florida. It is not always easy to separate these two bedbugs, but in *C. lectularius* the prothorax is generally 2.5 times as wide as long, whereas in *C. hemipterus* it is only about twice as wide as long. Also, in *C. hemipterus* the abdomen is not as rounded as in *C. lectularius*.

A third species, *Leptocimex boueti*, found only in West Africa, bites bats and also people, but is of much less importance than *Cimex* species.

Bedbugs are not considered important vectors, but in addition to constituting a biting nuisance they may aid the transmission of hepatitis B virus. In India bedbugs have been reported as causing iron deficiency in infants.

13.1 EXTERNAL MORPHOLOGY

Adult bedbugs are oval, wingless insects which are flattened dorsoventrally (Fig. 13.1). They are about 4–7 mm long and when unfed a pale yellow or brown colour, but after a full blood-meal they become a characteristically darker 'mahogany' brown. The head is short and broad and has a pair of prominent compound eyes, in front of which are a pair of four-segmented antennae. The proboscis is slender, and is normally held closely appressed along the ventral surface of the head and prothorax, but when the bug takes a blood-meal it is swung forward and downwards (Fig. 13.2*a*). The prothorax is much larger than the meso- and metathorax and

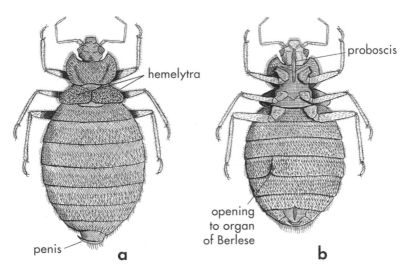

Figure 13.1 The bedbug, *Cimex hemipterus*. (*a*) Dorsal view of adult male; (*b*) ventral view of adult female.

has distinct wing-like expansions. The rudimentary and non-functional wings, termed the hemelytra, appear as two more or less oval pads overlying the meso- and metathorax. The three pairs of legs are slender but well developed.

The abdomen is divided into eight visible segments. In adult males the tip of the abdomen is slightly more pointed than in the females and careful examination shows that there is a small well-developed and curved penis (Fig. 13.1a). When viewed ventrally a small incision is seen on the left side of the fourth abdominal segment of females (Fig. 13.1b). This opens into a special pouch called the mesospermalege or the organ of Berlese or Ribaga, which serves to collect and store sperm. Since both male and female bugs bite people, it may not, however, be medically very important to distinguish the sexes.

13.2   LIFE CYCLE

Both sexes of bedbug take blood-meals and are equally important as pests. Feeding usually occurs on sleeping persons during the night, especially just before dawn, but if the bedbugs are starving they will feed during the day in dark rooms, or sometimes in light ones. Bedbugs, unlike lice, do not stay long on humans but visit them only to take blood-meals. In the absence of people bedbugs will feed on a variety of mammals, including rabbits, rats, mice and bats, and also on poultry and other birds. During the day both adults and nymphs are inactive and hide away in a variety of dark and dry places, such as cracks and crevices in furniture, walls, ceilings or floor-boards, underneath seams of wallpaper and between mattresses and beds. Bedbugs are gregarious and are frequently found in large numbers.

Females lay about 6–10 eggs a week which are deposited in the same places where the bugs hide, such as in cracks and crevices of buildings and furniture. The eggs are about 1 mm long, pearly or yellowish-white, covered with a very fine and delicate mosaic pattern, and characteristically slightly curved anteriorly (Fig. 13.2b). Some 500 or more eggs can sometimes be found more or less together, cemented on rough surfaces such as walls, or deposited in cracks. Females live several weeks to many months, and during this time may lay 50–200 eggs.

The eggs usually hatch after about 8–11 days, but within less than a week if temperatures are about 27 °C; if, however, temperatures in houses are low, hatching may be delayed for several weeks. At low temperatures eggs can survive for up to 3 months, but hatching will not occur below about 13 °C. During hatching the small operculum (cap) is pushed up from the anterior end of the egg, but often remains partially attached. Empty eggshells usually remain cemented in place after hatching. The newly hatched bedbug (nymph) is very pale yellow and resembles an adult, but is much smaller (Fig. 13.2c). The life cycle is hemimetabolous and there are

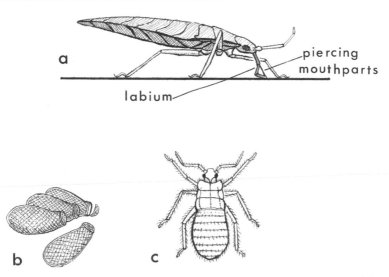

Figure 13.2 Bedbugs. (*a*) Diagram of adult with proboscis swung forward for feeding; (*b*) one hatched and three unhatched eggs; (*c*) first-instar nymph.

five nymphal instars, each of which takes one or more blood-meals. The nymphal period commonly lasts 5–8 weeks, but may be greatly extended in cool conditions or if regular blood-feeding is prevented by lack of hosts. The life cycle, from egg to adult (Fig. 13.3), can be as short as 3 weeks if temperatures are high and food plentiful, but more usually lasts 6 weeks to many months. Adults can, at least in the laboratory, live up to 4 years, and can withstand long periods of starvation – up to 550 days. Survival, however, is very much dependent on temperature and humidity.

The method of mating in bedbugs is unique among insects. The penis is not introduced into the genital opening but penetrates the integument and enters the organ of Berlese (organ of Ribaga or mesospermalege) situated on the ventral surface of the female. This organ serves as a copulatory pouch into which spermatozoa are introduced. After 1 or 2 hours spermatozoa leave this 'pouch' and pass into the haemocoel of the female, and then migrate to the bases of the oviducts and ascend to the ovaries where fertilization occurs.

Bedbugs in houses can be detected by the presence of live bugs, the cast-off skins of the nymphs, and hatched and unhatched eggs, all of which may be found in cracks and crevices. In addition, small dark brown or black marks may be visible on bed sheets, walls and wallpaper; these are the bedbug's excreta and consist mainly of excess blood ingested during feeding. Houses with heavy infestations of bedbugs may have a characteristic sweet and rather sickly smell, but in practice this may not be apparent because the weak odour is masked by stronger insanitary smells. At night male and female adults, and nymphs, crawl from their daytime resting

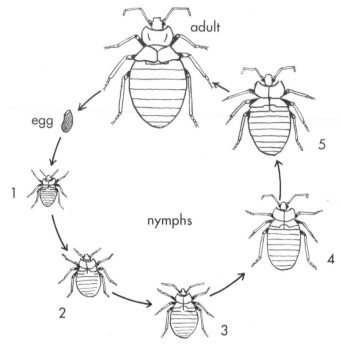

Figure 13.3 Life cycle of bedbugs.

places to feed on sleeping people, after which they return to their resting sites to digest their blood-meals. Bedbugs can move quite rapidly when disturbed.

Bedbugs have limited powers of dispersal because they do not have wings. Occasionally they may crawl from one building to another, but they usually spread to new houses by being introduced with furniture and bedding, or more rarely with clothing and hand baggage. Buying second-hand furniture can result in the introduction of bedbugs into houses.

13.3   MEDICAL IMPORTANCE

In the past bedbugs have not been considered to transmit disease to humans, but in Africa hepatitis B virus has been recorded from bedbugs. It has been suggested that the virus could be spread through contaminated faeces being scratched into abrasions, or even by inhalation of faecal dusts, but there was no evidence of bedbug involvement in a study in The Gambia.

In some areas, especially those with dilapidated buildings and low standards of hygiene, bedbug infestations can cause considerable distress. Reaction to their bites is variable. Some people show little or no reaction whereas others may suffer severe reactions and sleepless nights. In India repeated feedings of large numbers of bedbugs has been reported as responsible for anaemia in infants.

13.4   CONTROL

The use of insect repellents and pyrethroid-impregnated bed nets may afford considerable personal protection against bedbugs.

Floors and walls of infested houses together with as much furniture as possible should be sprayed with 5% DDT emulsion. If, however, bedbugs are resistant to DDT then 0.5% HCH should be tried, and if the bugs are also resistant to this compound, other insecticides such as 1–2% malathion, 0.5% diazinon, 0.5% dichlorvos (DDVP), 1% carbaryl (Sevin) or 2% propoxur (Baygon) can be used. The addition of 0.1–0.2% pyrethrins or synthetic pyrethroids to sprays is useful because it helps flush out bedbugs from their hiding places, which increases their contact with the insecticide. Pyrethroids, such as 0.5% permethrin, 0.005% deltamethrin or 0.005% lambdacyhalothrin (Icon), can also be used on their own to kill bedbugs.

Bedding and mattresses can be lightly sprayed with insecticides (not diazinon), but should be aired afterwards to allow them to dry out completely before being re-used. Insecticidal dusts can also be applied to mattresses and bedding. Infants' bedding should not be treated with insecticides.

Commercially available small insecticidal smoke generators containing, for example, DDT, HCH, permethrin or pirimiphos methyl (Actellic), which burn for up to 15 minutes, can be used to fumigate infested premises.

FURTHER READING

Anon. (1973) *The Bedbug*, 8th edn. Economic Series 5, 16pp. London: British Museum (Natural History).

Johnson, C.G. (1941) The ecology of the bedbug, *Cimex lectularius* L., in Britain. *Journal of Hygiene, Cambridge*, **41**, 345–461.

King, F. (1990) Mind the bugs don't bite. *New Scientist*, 27 January, 51–4.

Mayans, M.V., Hall, A.J., Inskip, H.M., Lindsay, S.W., Chotard, J., Mendy, M. and Whittle, H.C. (1994) Do bedbugs transmit hepatitis B? *Lancet*, **343**, 761–3.

Ryckman, R.E., Bentley, D.G. and Archbold, E.F. (1981) The Cimicidae of the Americas and Oceanic Islands, a checklist and bibliography. *Bulletin of the Society of Vector Ecologists*, **6**, 93–142.

Usinger, R.L. (1966) *Monograph of Cimicidae (Hemiptera–Heteroptera)*. College Park, Maryland: Entomological Society of America.

Venkatachalam, P.S. and Belavady, B. (1962) Loss of haemoglobin iron due to excessive biting by bed bugs. A possible aetiological factor in the iron deficiency anaemia of infants and children. *Transactions of the Royal Society of Tropical Medicine and Hygiene*, **56**, 218–21.

Weidhaas, D.E. and Keiding, J. (1982) VI. Bed bugs. WHO/VBC/82.857. Geneva: World Health Organization mimeographed document.

14

Triatomine bugs (Triatominae)

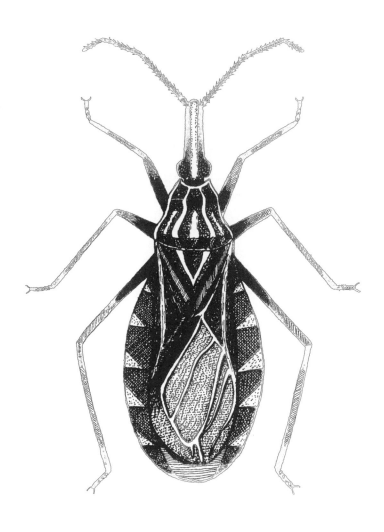

The blood-sucking bugs of the family Reduviidae belong to the subfamily Triatominae and comprise 126 species in 15–17 genera (the number depending on whose authority is accepted). Principal species of medical importance are *Triatoma infestans*, *T. dimidiata*, *T. brasiliensis*, *Rhodnius prolixus* and *Panstrongylus megistus*, all of which spread Chagas disease (*Trypanosoma cruzi*) in Central and South America. Some can also transmit *Trypanosoma rangeli*, an apparently non-pathogenic organism.

Most Triatominae occur in the Americas, ranging from the Great Lakes of the USA to southern Argentina, but 13 species are found in the Old World tropics. All medically important species are confined to the southern USA, Central and South America. Triatomines are commonly called kissing-bugs or cone-nose bugs.

14.1 EXTERNAL MORPHOLOGY

Triatominae vary in size from about 0.5–4.5 cm, but most are 2–3 cm long. They are easily recognized by their long snout-like head which bears a pair of prominent dark-coloured eyes, in front of which are a pair of laterally situated, long and thin four-segmented antennae (Fig. 14.1). The proboscis, sometimes called the rostrum, is relatively thin and straight and, as in bedbugs, lies closely appressed to the ventral surface of the head (Fig. 14.2*a*). When the Triatominae take a blood-meal the proboscis is swung forward and downwards (Fig. 14.2*b*). Bugs in other subfamilies of the

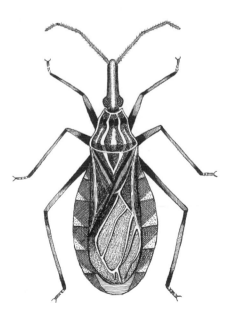

Figure 14.1 Adult *Rhodnius*, as an example of a triatomine bug.

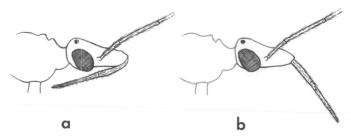

a   b

Figure 14.2 Lateral view of the head of *Triatoma*. (*a*) Proboscis closely appressed to ventral side of head; (*b*) proboscis swung forwards in feeding position.

Reduviidae are predacious on small insects and the proboscis is usually thicker and more robust than in the Triatominae, and is sometimes distinctly curved. Another difference is that the antennae arise from the dorsal surface of the head and not from its sides as in the blood-sucking Triatominae (Fig. 14.2).

The dorsal part of the first segment of the thorax of the Triatominae consists of a very conspicuous triangular pronotum. The meso- and metathorax are completely hidden dorsally by the folded fore-wings, which are called hemelytra. The basal part of each hemelytron is thickened and relatively hard, whereas the more distal part is membranous (Fig. 14.3). The hind-wings are entirely membranous but when the bug is not flying they remain hidden underneath the hemelytra. The thorax has three pairs of relatively long and slender legs which end in paired small claws.

The abdomen is more or less oval in shape but is mostly covered by the wings, except for the lateral margins which are bent upwards slightly and are visible dorsally.

Triatominae are frequently a rather dull brown–black colour, but some species are more colourful, having contrasting yellow, orange, pink or red markings, usually as bands on the pronotum, basal part of the fore-wings, legs or margins of the abdomen.

basal part   distal part

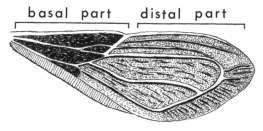

Figure 14.3 Fore-wing (hemelytron) of a triatomine bug showing thickened basal part and more membranous distal part.

14.2   LIFE CYCLE

Eggs are deposited in or near the habitation of their hosts, such as in cracks and crevices in walls, floors, ceilings and furniture of houses, especially dilapidated mud-walled and thatched-roofed houses in rural areas, or slums at the edge of towns. Adults of some species also lay their eggs in rodent burrows and a variety of shelters used by mammalian hosts upon which the bugs feed. Other species feed on birds and deposit their eggs in birds' nests and on leaves of trees. Eggs are about 1.5–2.5 mm long, pearly white, pink or yellowish depending on the species. They have a smooth or ornate shell and are oval in shape, but have a slight constriction before the operculum (cap) (Fig. 14.4). Some species lay eggs in small batches which may be glued to the substrate; others lay them more or less singly, either free or cemented to the substrate. The total number of eggs laid by females varies between about 50 and 1000 but is usually 200–300, depending on the species, their longevity and the number of blood-meals they take. The life cycle is hemimetabolous.

Small pale nymphs, which resemble adults but lack wings, may hatch from the eggs after only 10–15 days, but the incubation period may extend to 30, or even 60, days. These newly emerged nymphs usually remain hidden in cracks and crevices for a few days before they seek out blood-meals. There are five nymphal instars, each instar requiring at least one complete blood-meal before it changes into a succeeding one. Rudimentary wing pads begin to be clearly visible in the fourth and fifth nymphal stages, but only adults have fully developed functional wings. Young nymphs can ingest as much as 6 times their own weight of blood, and as a result their abdomens may become so greatly distended that they resemble blood-red balloons. Successive instars take relatively less blood, so that the fifth and last nymphal stage takes about 3–5 times its own weight of blood, and adults ingest 2–4 times their weight of blood. The largest bug, *Dipetalogaster maxima*, can take up more than 4 mg of blood. Sometimes hungry nymphs and adult bugs pierce the swollen abdomens of freshly engorged nymphs and take a blood-meal from them without apparently causing any harm.

Nymphs and adults of both sexes feed at night on their hosts, and

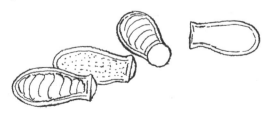

Figure 14.4 Hatched and unhatched triatomine eggs.

feeding is a lengthy process lasting 10–25 minutes. When people are covered with blankets the bugs feed on exposed parts of the body such as the nose and around the eyes and mouth, but they will readily feed on other exposed areas of the body. Bites are usually relatively painless and do not awaken people, but some species cause considerable discomfort and there may be prolonged after-effects. Many Triatominae defaecate during or soon after feeding and this behaviour is very important in the transmission of Chagas disease. The presence of bugs in houses is often characterized by streaks of black and white faeces on walls and furniture.

Because of the relatively long time required, even under optimum laboratory conditions, to digest their large blood-meals, the life cycle from egg to adult takes at least 3–4 months but more usually 6–10 months. Under natural conditions it often takes about 1 year, but sometimes the life cycle is 2 years. In the absence of hosts, older nymphs and adults can survive 4–6 months of starvation.

Species of bugs in the subfamily Triatominae inhabit both forests and drier areas of the Americas. Many species feed on a variety of wild animals, such as armadillos, opossums, rats, mice, marsupials, ground squirrels, skunks, iguanas, bats, and also birds. Adults and nymphs are usually found in the burrows or nests of these animals. In addition to these sylvatic species, certain bugs have become highly domesticated and feed on animals such as donkeys, cattle, goats, horses, pigs, cats, dogs and especially chickens, which in some areas appear to be particularly important hosts, and of course humans. These domestic species often live in man-made shelters including houses, especially primitive ones made of wood, mud and thatch. Some species are partially sylvatic and domestic in their feeding and resting habits. Sylvatic species sometimes readily move into houses from the forest as it is cut down and people occupy previously uninhabited areas.

If hosts vacate their shelters or homes, hungry nymphs eventually crawl out from their hiding places to seek new hosts, whereas adults, which are strong fliers, fly out to find new hosts and shelters. Some species are attracted into houses by lights.

14.3   MEDICAL IMPORTANCE

14.3.1 Chagas disease
In rural areas of the Americas there may be hundreds of triatomine bugs in a house and such infestations can be very stressful to the occupiers, who will receive many bites during the night. Large bug populations can contribute to anaemia. But medically the main importance of the Triatominae is that they transmit *Trypanosoma cruzi*, the causative agent of Chagas disease, sometimes referred to as South American trypanosomiasis. Parasites ingested with a blood-meal undergo their entire development

within the gut of the bug. After 9–17 days, sometimes longer, infective metacyclic trypomastigotes of *T. cruzi* are present in the lumen of the hindgut. Bugs may also become infected by feeding on recently engorged nymphs. Blood-feeding commonly lasts 10–25 minutes or longer, and during this time, or soon afterwards, many species of bugs excrete liquid or semiliquid faeces which may be contaminated with the metacyclic forms of *T. cruzi* derived from a previous blood-meal. People become infected when the excreta is scratched either into abrasions in the skin or in the site of the bug's bite, or when it gets rubbed into the eyes or other mucous membranes. If the bug's bite produces local irritation causing the person to scratch, this facilitates infection. Transmission is not by the bite of the insect, solely through its faeces.

It appears that all Triatominae of the Western hemisphere can transmit Chagas disease, and about 70 species have been recorded naturally infected; but some strains of *T. cruzi* are unable to develop in the gut of some species. In practice, however, only about a dozen species living in very close association with people and, therefore regularly feeding on humans, are principal vectors, and of these the most important are *Triatoma infestans*, *Panstrongylus megistus*, *Rhodnius prolixus*, *T. brasiliensis* and *T. dimidiata*. The efficiency of a vector will depend on its speed of feeding and whether or not it defaecates on a person during feeding.

Chagas disease is a zoonosis. *Trypanosoma cruzi* is essentially a parasite of wild animals, such as opossums (especially *Didelphis* species), armadillos, many species of wild and urban rats and mice, squirrels, carnivores, monkeys and possibly bats, all of which may serve as reservoirs of infection for humans. The bug itself is also a reservoir of infection, but in some areas humans are considered to be the principal one. Apart from acquiring infections with *T. cruzi* through faeces, some animals become infected by eating the bugs, or by eating infected animals, such as carnivores eating rodents infected with *T. cruzi*. People can also become infected by eating infected meat (e.g. inadequately cooked opossums) or food contaminated with excrement of infected bugs.

The infection rates of Triatominae are often exceptionally high. For example, it is not uncommon to find infection rates of about 25% or even 40% or more. Even higher infection rates (78%) have been found in *Triatoma protracta* in California, but because this species very rarely bites people it is not considered a vector to humans.

14.3.2 *Trypanosoma rangeli*

Another trypanosome, *Trypanosoma rangeli*, which is apparently non-pathogenic in humans, is also transmitted by the Triatominae. In the insect vector the trypanosomes undergo dual development, some or the metacyclic infective forms migrating to the hind-gut while others penetrate the gut wall, pass across into the haemocoel and then migrate to the salivary glands. Humans can be infected by both the bug's faeces and bite, but the latter seems to be the more important method of transmission, and some have questioned whether infection ever arises from the bug's faeces.

14.4   CONTROL

The usual control methods consist of applying residual insecticides to the interior surfaces of walls and roofs of houses. For many years 500 mg/m^2 of HCH was used, but this has been replaced by the pyrethroids, particularly deltamethrin (25 mg/m^2), cyfluthrin (50 mg/m^2) and lambdacyhalothrin (Icon) (30 mg/m^2), and to a lesser extent by cypermethrin (125 mg/m^2). A single spraying should eliminate populations of domestic bugs, and may sometimes also eliminate peridomestic ones. Careful surveillance is then carried out to detect any reinfestation or foci of bugs that need further spraying.

Slow-release formulations of insecticidal paints based on latex or polyvinyl acetate can be painted or sprayed onto most indoor surfaces, including mud walls. Malathion or fenitrothion (Sumithion) are usually the insecticides incorporated in these paints. They can remain effective for up to 2 years. The wooden framework of new mud-and-wattle houses can be painted with bitumen-type paint containing malathion or chlorpyrifos (Dursban) before the mud is applied. Such treated framework remains toxic for at least 5 years. However, this control method is not widely used because of difficulties in applying the paints and resins on a large scale.

The above methods will destroy bugs resting in houses but are less effective against those resting in natural outdoor shelters. These peridomestic populations can sometimes reinstate domestic populations.

Additional methods of control involve making houses unattractive as resting sites for the bugs, for example by plastering walls to cover up cracks in which the bugs might live, and replacing dilapidated mud and thatched houses with those built of bricks or cement blocks and having corrugated metal roofs. However, because of the high costs of building new houses, rehousing has yet to be carried out on a large scale.

14.4.1   'Southern cone initiative'

In 1991 Argentina, Bolivia, Brazil, Chile, Paraguay and Uruguay proposed a plan of action to eliminate Chagas disease through a combination of vector control methods, mainly insecticidal spraying of houses, and a blood screening programme. In 1997 Uruguay was declared free of Chagas disease transmission. In the same year two new initiatives were launched to interrupt transmission in the Central American countries of Belize, Costa Rica, El Salvador, Guatemala, Honduras, Nicaragua and Panama, and also in the Andean countries of Colombia, Venezuela and Ecuador.

FURTHER READING

Barrett, T.V. (1991) Advances in triatomine bug ecology in relation to Chagas' disease. *Advances in Disease Vector Research*, **8**, 143–76.

Brenner, R.R. and Stoka, A. de la M. (eds.) (1988) *Chagas' Disease Vectors*. I. *Taxonomic, Ecological and Epidemiological Aspects*. Boca Raton, Florida: CRC Press.

Brenner, R.R. and Stoka, A. de la M. (eds.) (1988) *Chagas' Disease Vectors*. II. *Anatomic and Physiological Aspects*. Boca Raton, Florida: CRC Press.

Brenner, R.R. and Stoka, A. de la M. (eds.) (1988) *Chagas Disease Vectors*, III. *Biochemical Aspects and Control*. Boca Raton, Florida: CRC Press.

Briceno-Leon, R. (1987) Rural housing for control of Chagas' disease in Venezuela. *Parasitology Today*, **3**, 384–7.

Bryan, R.T., Balderrama, F., Tonn, R.J. and Dias, J.C.P. (1994) Community participation in vector control: lessons from Chagas' disease. *American Journal of Tropical Medicine and Hygiene*, **50** (Suppl.), 61–71.

Dias, J.C.P. (1987) Control of Chagas' disease in Brazil. *Parasitology Today*, **3**, 336–41.

Dias, J.C.P. (1988) Controle de vetores da doença de Chagas no Brasil e riscos da reinvasão domiciliar por vetores secundários. *Memórias do Instituto Oswaldo Cruz*, **83** (Suppl. I, Special Issue), 387–91.

Forattini, O.P. (1989) Chagas' disease and human behavior. In *Demography and Vector-Borne Diseases*, ed. M.W. Service, pp. 107–20. Boca Raton, Florida: CRC Press.

Kingman, S. (1991) South America declares war on Chagas' disease. *New Scientist*, 19 October, 16–17.

Schofield, C.J. (1979) The behaviour of Triatominae (Hemiptera: Reduviidae): a review. *Bulletin of Entomological Research*, **69**, 363–79.

Schofield, C.J. (1988) Biosystematics of the Triatominae. In *Biosystematics of Haematophagous Insects*, ed. M.W. Service, pp. 280–312. Oxford: Clarendon Press.

Schofield, C.J. (1994) *Triatominae. Biology and Control*. Bognor Regis: Eurocommunica Publications.

Schofield, C.J. and Dias, J.C.P. (1998) The Southern Cone Initiative against Chagas disease. *Advances in Parasitology*, **42**, 1–27.

Schofield, C.J. and Dujardin, J.P. (1997) Chagas disease vector control in Central America. *Parasitology Today*, **13**, 141–4.

15

Cockroaches (Blattaria)

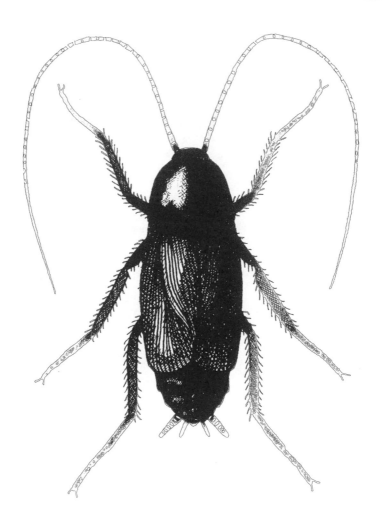

There are almost 4000 species of cockroaches. About 50 species have become domestic pests, and the most important medically are *Blattella germanica* (the German cockroach), *Blatta orientalis* (the Oriental cockroach) and *Periplaneta americana* (the American cockroach). Cockroaches are sometimes called roaches or steambugs. They have an almost world-wide distribution. They belong to the suborder Blattaria of the order Dictyoptera.

Cockroaches almost certainly aid in the transmission and harbourage of various pathogenic viruses, bacteria, protozoans and helminths. They can be intermediate hosts of certain nematodes, and an acanthocephalid which rarely parasitizes humans.

15.1 EXTERNAL MORPHOLOGY

The following general description refers to the more common household pest species. Cockroaches are usually chestnut brown or black, 1–5 cm long, flattened dorsoventrally and have a smooth, shiny and tough integument. Viewed from above the head appears small, and it is sometimes almost hidden by the large, rounded pronotum. A pair of long and prominent filiform antennae arise from the front of the head between the eyes (Fig. 15.1). The cockroach mouthparts are developed for chewing, gnawing and scraping; they are not composed of piercing stylets and therefore cockroaches cannot suck blood. In adults of both sexes there are two pairs of wings. In certain household species those in the female may be shorter than those of the male, and in female *Blatta orientalis* they are very small and non-functional. The fore-wings are thickened and rather leathery in

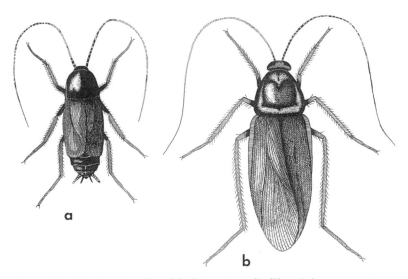

Figure 15.1 Adult cockroaches. (*a*) *Blatta orientalis*; (*b*) *Periplaneta americana*.

texture and are called tegmina; they are not used in flight but serve as protective covers for the hind-wings, which are membranous and can be used for flying, but when not in use are folded shut, fan-like over the body. Although cockroaches possess wings they rarely fly in temperate climates. There are three pairs of legs which are well developed and covered with prominent small spines and bristles; the five-segmented tarsi end in a pair of claws.

The segmented abdomen is more or less oval in shape but is either completely or partly hidden from view, depending on the species, by the folded overlapping wings. In both sexes a pair of prominent segmented pilose cerci arise from the last abdominal segment (Fig. 15.2a), but they are hidden from view in some species by the wings.

Cockroaches are most readily distinguished from beetles (order Coleoptera) by having the fore-wings placed over the abdomen in a scissor-like manner. In beetles the fore-wings (elytra) are not crossed over but meet dorsally to form a distinct line down the centre of the abdomen. In addition, the elytra of beetles are generally stouter than the tegmina of cockroaches.

15.2   LIFE CYCLE

Cockroaches like warmth and during the day they hide away behind radiators and hot-water pipes, and in warm countries where these may be absent, in almost any dark place, such as cesspits, septic tanks, sewers, rubbish dumps, refuse tips, dustbins, cupboards, drawers, underneath chairs, tables, sinks, baths and beds, behind refrigerators, cooking stoves, dishes in kitchens. They are usually common in kitchens, especially if remains of food are left out overnight. They abound in restaurants, hotels, bakeries, breweries, laundries and aboard ships. They are also found throughout the world in hospitals. Cockroaches are nocturnal in habit and are rarely seen during the day unless they are disturbed from their hiding places. They become very active at night, crawling over floors, tables and other furniture to seek food. They move very quickly and can be both seen and heard scuttling along when lights are suddenly switched on.

They are omnivorous and voracious feeders; any type of food is eaten. They also consume paper, clothes, particularly starched ones, books, hair, shoes, wallpaper, dried blood, sputum, excreta, dead insects, and almost any animal or vegetable matter. They have been recorded gnawing at the finger- and toe-nails of babies or sleeping or comatose people, and even infesting the hair of vagrants. Cockroaches habitually disgorge partially digested meals and deposit their excreta on almost anything, including food. They may live for 5–10 weeks without water and for many months without food, but in practice this is not an important limiting factor

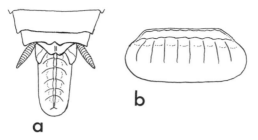

Figure 15.2 (a) Ootheca protruding from the abdomen of *Blatella germanica*; (b) typical cockroach ootheca, lateral view.

because they very rarely occur in areas where food of some kind is not available. Young nymphs, however, may die within about 7–10 days in the absence of food.

Eggs are laid encased in a brown bean-shaped case or capsule called an ootheca (Fig. 15.2*b*), which can contain 12–50 eggs but usually 24–40. Cockroaches, especially *Blattella germanica*, are often seen running around with an ootheca partly protruding from the tip of the abdomen (Fig. 15.2*a*). The oothecae are deposited in cracks and crevices in dark and secluded places. In some species they are cemented to surfaces such as the undersides of tables, chairs and beds.

Adult cockroaches live for many months to 2 years or more, and during this time the female will lay about 4–90 oothecae, the number varying considerably according to species. Cockroaches have a hemimetabolous life cycle. Nymphs hatch from the eggs after about 1–3 months, the time depending on both temperature and species. Young nymphs are very pale and delicate versions of the adults, whereas the older ones are progressively darker and resemble the adults more. The nymphs are wingless, the wings gradually develop with ensuing nymphal instars, but only adults may have fully developed wings. There are usually six nymphal instars, but in *Periplaneta americana* there may be as many as 13 nymphal instars. The duration of the nymphal stage varies according to temperature, abundance of food and species. For example, the nymphal stage may occupy only about 2–3 months in *Blattella germanica*, 12–15 months in *Blatta orientalis*, and up to 10–23 months in *Periplaneta americana*. Under less ideal conditions the nymphal period of *Blattella germanica* extends to 6 months, and up to about 3 years in *Periplaneta americana*.

Cockroaches spread very rapidly from infested houses to adjoining ones. They often gain entry by climbing up water pipes and waste pipes. They are also spread as oothecae, nymphs and adults in furniture and other belongings.

15.3   MEDICAL IMPORTANCE

The presence of cockroaches in houses and hotels, etc. has for a long time been regarded as highly undesirable because of their dirty habits of feeding indiscriminately on both excreta and foods, and their practice of excreting and regurgitating their partially digested meals over food.

Many of the parasites, and pathogens, isolated from cockroaches are spread directly from person to person without the aid of intermediary insects, and because of this it is very difficult to prove that cockroaches are the prime cause of any disease outbreak. Nevertheless because of their insanitary habits they have been suspected as aiding the transmission of various illnesses. For example, they are known to carry pathogenic viruses such as poliomyelitis, protozoans such as *Entamoeba histolytica*, *Trichomonas hominis*, *Giardia intestinalis* and *Balantidium coli*, and bacteria such as *Escherichia coli*, *Staphylococcus aureus*, *Klebsiella pneumoniae*, *Shigella dysentariae* and *Salmonella* species, including *S. typhi* and *S. typhimurium*. They are also known to be intermediate hosts of the acanthocephalid, *Moniliformis moniliformis*, which is common in rodents and can also infect cats and dogs, and very rarely people. Nematodes such as *Gonglyonema pulchrum*, a common parasite of herbivores and occasionally humans, and *Enterobius vermicularis*, which is an extremely common worm in humans, can also be carried by cockroaches. They have also been found naturally infected with *Toxoplasma gondii* and suspected of transmitting this parasite to cats, and possibly to people, by feeding on cats' faeces.

There is little doubt that cockroaches contribute to the spread of a number of diseases, mainly intestinal, and they may sometimes be more important as mechanical vectors than house-flies. However, it is nevertheless difficult to assess their real importance as vectors, because many of the pathogens which cockroaches carry can be transmitted by many other different ways.

Some people are allergic to cockroaches. It appears that sensitized people can react to cockroach allergens by eating cockroach-contaminated food, or by inhaling their dried faecal pellets.

15.4   CONTROL

Ensuring that neither food nor dirty kitchen utensils are left out overnight will help reduce the number of cockroaches, but if they are present in adjoining or nearby houses, good hygiene in itself will not prevent cockroaches from entering houses.

Insecticidal spraying or dusting of selected sites such as cupboards, wardrobes, kitchen furniture and fixtures, underneath sinks, stoves, refrigerators, and nearby dustbins, is recommended. Organochlorine resistance is common in many pest species. In some areas *Periplaneta americana* is also resistant to the organophosphates, and *Blattella germanica* is often resistant to the organophosphates, carbamates and pyrethroids. Insect growth regulators (IGRs) have more recently been used to control cockroaches when they have developed resistance to a broad spectrum of insecticides, but their use is hindered by high costs and limited availability.

In the absence of resistance 0.5–3% sprays or 0.5–5% dusts of organophosphate insecticides such as pirimiphos methyl (Actellic), fenitrothion (Sumithion), malathion, diazinon, chlorpyrifos (Dursban) or dichlorvos (DDVP), or carbamates such as propoxur (Baygon), carbaryl (Sevin) or bendiocarb (Ficam) can be used. Sprays based on kerosene (paraffin) may leave unsightly stains on walls, and are of course dangerous near naked flames or cookers.

The residual efficiency of sprays greatly depends on the surfaces on which they are applied; for example on most painted and shiny surfaces residual activity often lasts only about 1–4 weeks. The residual action of insecticidal dusts is less affected by the nature of the surface, but dusting is unsightly and therefore objected to by many householders. The addition of pyrethrum or pyrethroids to insecticidal sprays or aerosols is useful because it irritates cockroaches and flushes them out from their hiding places.

Pyrethroids such as permethrin, deltamethrin, lambdacyhalothrin (Icon) of cyfluthrin applied as sprays, or more effectively as aerosols, can produce spectacular results in both flushing out cockroaches and killing them. Mixtures of an organophosphate or carbamate insecticide and a pyrethroid can be very effective in not just flushing out cockroaches, but also giving a rapid kill and long-term insecticidal activity.

Various commercial lacquers and varnishes containing residual insecticides such as 1% cypermethrin, 0.5% alphacypermethrin or 2% diazinon can be painted onto walls and other surfaces and remain effective for several months in killing cockroaches.

Boric acid powder (borax) still remains a very safe and useful chemical, acting both as a contact insecticide and a stomach poison.

Most organophosphate and carbamate insecticides can be added (1–2%) to baits of food, such as peanut butter, dog food and maltose, to which glycerol may be added to increase their attractiveness. Solid baits can be placed in areas known to harbour large numbers of cockroaches; when eaten they cause their death. Alternatively attractant baits or pheromones can be placed in various simple cardboard or sticky traps to entice cockroaches into them, after which they are killed or prevented from escaping.

Another approach to reduce cockroach infestations is to apply repellents such as dimethylphthalate (dimp) or diethyltoluamide (deet) to kitchen furniture and storerooms, etc.

FURTHER READING

Burgess, N.R.H. (1984) Hospital design and cockroach control. *Transactions of the Royal Society of Tropical Medicine and Hygiene*, **78**, 293–4.

Burgess, N.R.H. and Chetwyn, K.N. (1981) Association of cockroaches with an outbreak of dysentery. *Transactions of the Royal Society of Tropical Medicine and Hygiene*, **75**, 332–3.

Cochran, D.G. (1982) V. Cockroaches: biology and control. WHO/VBC/82.856. Geneva: World Health Organization mimeographed document.

Cornwell, P.B. (1968). *The Cockroach*, vol. 1, *A Laboratory Insect and an Industrial Pest*. London: Hutchinson.

Cornwell, P.B. (1976) *The Cockroach*, vol. 2, *Insecticides and Cockroach Control*. London: Associated Business Programmes.

Guthrie, D.M. and Tindall, A.R. (1968) *The Biology of the Cockroach*. London: Edward Arnold.

Rust, M.K., Reierson, D.A. and Hangsen, K.H. (1991) Control of American cockroaches (Dictyoptera: Blattidae) in sewers. *Journal of Medical Entomology*, **28**, 210–3.

Schal, C. and Hamilton, R.L. (1990) Integrated suppression of synanthropic cockroaches. *Annual Review of Entomology*, **35**, 521–51.

16

Soft ticks (Argasidae)

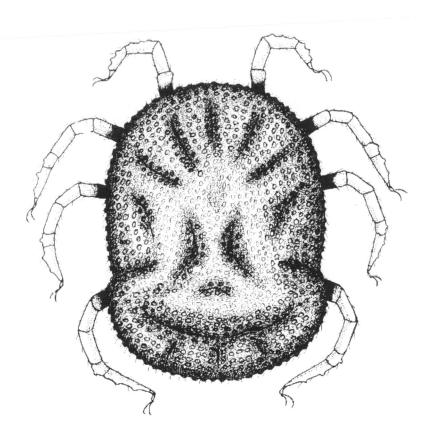

Ticks, mites, spiders and scorpions belong to the class Arachnida, whereas the subclass Acari comprises just the ticks and mites; but the most important point to remember is that ticks are divided into two main families, namely the Argasidae (soft ticks) and the Ixodidae (hard ticks). (A third family (Nuttalliellidae) contains just one obscure species which is of no medical importance.) Another important point is that ticks differ from mites in being much larger, although their small immature stages may approach the size of some mites, but ticks differ from mites in having a toothed hypostome (Fig. 16.2) as well as by not having claws on the palps.

Soft ticks (Argasidae) have a more or less world-wide distribution. There are some 180 species belonging to 11 genera, but the medically important soft ticks belong to the genus *Ornithodoros*. The most important vector is *Ornithodoros moubata*, which is a species within the *O. moubata* complex. The precise relationships of the six species and subspecies in this complex are still unresolved. *Ornithodoros moubata* is the main vector of tick-borne (endemic) relapsing fever, which is caused by *Borrelia duttoni*.

16.1   EXTERNAL MORPHOLOGY

Adult argasid ticks are flattened dorsoventrally and oval in outline, but the shape varies according to the species. The integument is tough and leathery, wrinkled and usually has fine tubercles (mammillae) or granulations. There is no scutum or dorsal shield as is found in ixodid (hard) ticks. The mouthparts termed the capitulum, gnathosoma, or 'false head', are situated ventrally (Fig. 16.1b) and thus not visible dorsally. This character serves to separate adults and nymphal soft ticks from hard ticks (Ixodidae), which have the capitulum projecting forward and clearly visible dorsally. The four-segmented palps are leg-like and the chelicerae have smooth, not denticulate, sheaths (Fig. 16.2). The powerful cutting chelicerae have strong teeth at their tips and together with the hypostome, which has teeth arranged in several longitudinal rows, penetrate the host during feeding.

The four pairs of legs are well developed and terminate in a pair of claws. The coxal organs (glands) open between the bases of the coxae of the first and second pairs of legs (Fig. 16.1b), and are osmoregulatory in function. Spiracles (stigmata) are paired structures located on the margin of the body (idiosoma) just behind the last pair of legs and are often surrounded by circular, oval or comma-shaped plates. The genital opening (vulva) is situated ventrally just behind the capitulum.

Males and females are very similar in appearance and usually difficult to separate, although blood-engorged females can be considerably larger than males because they ingest much greater amounts of blood. However, as both sexes feed on blood and can consequently be disease vectors, it is not so important to be able to distinguish between them.

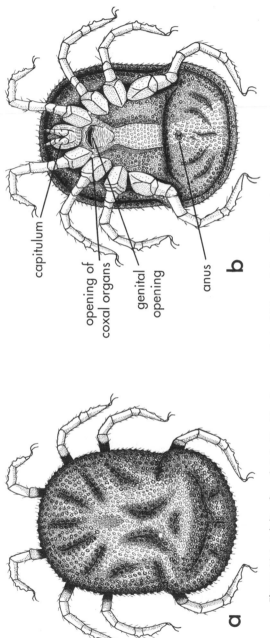

Figure 16.1 Adults of a soft tick, *Ornithodoros moubata*. (*a*) Dorsal view; (*b*) ventral view.

capitulum

opening of coxal organs

genital opening

anus

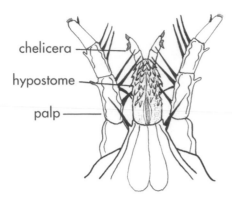

chelicera

hypostome

palp

Figure 16.2 Capitulum of an adult *Ornithodoros* species showing leg-like palps and non-denticulate cheliceral sheaths.

Medically the most important genus of argasids is *Ornithodoros*, species of which are found in many areas of the world including the Americas, Africa, Europe and Asia. The most important disease vector is *Ornithodoros moubata*.

16.2   INTERNAL ANATOMY

A brief account of the internal anatomy of a tick is necessary in order to understand the mechanisms of disease transmission.

During feeding, saliva, which often contains powerful anticoagulants, is secreted by a pair of large grape-like salivary glands and flows down the mouthparts. Blood from the host then passes up the mouthparts, through the narrow oesophagus and into the stomach or mid-gut, which is provided with numerous branching diverticula. The ramifications of the diverticula enable the adult tick to ingest large volumes of blood (about 6–12 times its own weight), resulting in great distension of the tick's body.

Argasid ticks have a pair of coxal organs, which although sometimes called coxal glands, are not actually glandular but filter off excess fluid and salts from ingested blood-meals. This fluid is passed out through a small opening located between the bases of the first two pairs of legs. When a soft tick is infected with tick-borne relapsing fever (*Borrelia duttoni*) many of the spirochaetes in the haemolymph enter the coxal organs and are then passed out through their openings.

Coxal organs are present only in soft ticks, not in hard ticks.

In females of both soft and hard ticks a peculiar structure termed Gene's organ is located in front of the mid-gut, and during oviposition is extruded from a small opening above the capitulum. It secretes waxy waterproofing substances which coat the eggs during oviposition and thus enables them

to withstand desiccation, immersion in water and other adverse environmental conditions.

16.3   LIFE CYCLE

A blood-meal is essential for maturation of the ovaries and egg laying.
Most female ticks ingest large blood-meals; soft ticks may increase their
weight 12-fold after feeding, and hard ticks take up even more blood (p.
230). After each blood-meal female argasid ticks lay several (often 4–6)
small egg batches, each of about 15–100 spherical eggs. Occasionally fewer
but larger egg batches comprising as many as 300–500 eggs are laid. Adult
ticks can live for many years, so a female may lay thousands of eggs during
her lifetime. Eggs are deposited in or near the resting places of the adult
ticks, such as in cracks and crevices in the walls, floors and furniture of
houses, or in mud, dust and debris, in rodent holes or in the more exposed
resting or sleeping places of wild animals and birds.

Eggs hatch usually within 1–3 weeks, but because they have been coated
during oviposition with a protective waxy secretion from Gene's organ
(see above) they can remain viable for many months under adverse climatic conditions.

Both argasid and ixodid ticks have a hemimetabolous life cycle, that is
eggs hatch to produce six-legged larvae which superficially resemble the
adults, and which moult to produce eight-legged nymphs which resemble
the adults even more closely. In argasid ticks the six-legged larva (Fig. 16.3)
is usually very active and searches for a host from which to take a blood-
meal. The capitulum projects from the body and is visible from above.

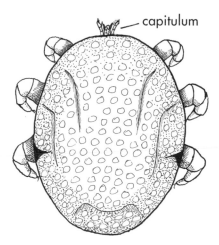

Figure 16.3 Larva of *Ornithodoros* species, as an example of a typical soft tick
showing capitulum projecting in front of the body.

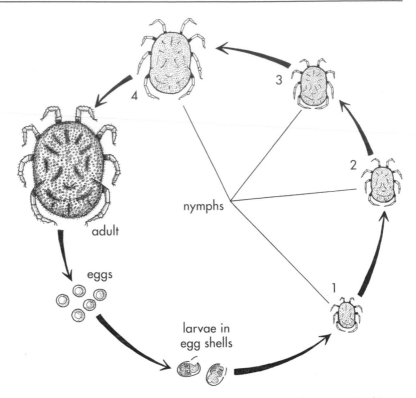

Figure 16.4 Life cycle of *Ornithodoros moubata* showing larvae retained in eggshells and four nymphal stages.

Blood-feeding on the host typically lasts for 20–30 minutes, but is sometimes completed in 3–4 minutes, after which the engorged larva drops to the ground and after a few days moults to produce an eight-legged nymph. The nymph seeks out a host and feeds for about 20–30 minutes before it falls to the ground. Argasid ticks usually have four or five nymphal instars (Fig. 16.4), but up to seven in some species. Each nymphal stage requires a blood-meal before it can proceed to the next nymphal instar.

Larvae of *Ornithodoros moubata* differ from most other argasid ticks in that they do not take blood-meals but remain within their eggshells after hatching, moulting to produce first-instar nymphs which crawl from the eggshells to seek their blood-meals.

The duration of the life cycle, from egg hatching to adult, depends on the species of tick, temperature and the availability of blood-meals, but is typically about 6–12 months. Adult ticks can live for many years, up to 12–20 years in the laboratory; the record for a tick is 25 years! In the absence of suitable hosts adults can survive up to 12 years without a blood-meal. In argasid ticks mating usually occurs away from the host, such as on the ground or amongst vegetation.

The distribution of the larvae, nymphs and adults of argasid ticks is usually patchy and restricted to the homes, nests and resting places of their hosts, but in these places there can be quite large populations of ticks. Species which commonly feed on people, such as *O. moubata* in Africa, are found around human settlements, especially in village huts. They can also be found in livestock shelters, chicken sheds, animal burrows and caves, especially in dry areas. However, in many parts of Africa these ticks appear to be becoming uncommon. This may be due to changes in life-style, in particular to the increased numbers of people sleeping on beds raised from the floor, which makes it more difficult for the ticks to feed on humans.

Because the larvae and all nymphal instars and adults take blood, but nevertheless remain attached for only relatively short periods, many hosts, comprising both different species and individuals, are fed upon during their life cycle. Argasid ticks are consequently referred to as 'many-host' or 'multi-host' ticks.

16.4   MEDICAL IMPORTANCE

16.4.1 Tick-borne relapsing fever

Soft ticks can inflict painful bites and although they can sometimes cause tick paralysis this condition is more common following prolonged feeding by hard ticks (p. 232). Tick-borne relapsing fever is the only important disease transmitted to humans by soft ticks. The infection occurs throughout most of the tropics, subtropics and in many areas of the temperate region such as North America and Europe, but is absent from Australia and New Zealand.

There are about seven species of *Borrelia* each having different geographical distributions that cause *Ornithodoros*-transmitted relapsing fever, but the most common is *B. duttoni*, which is spread by *O. moubata* in sub-Saharan Africa.

Spirochaetes ingested with a blood-meal multiply in the gut and congregate along the wall of the tick's mid-gut and then pass across into the haemocoel where they can be found after 24 hours. In the haemocoel, the spirochaetes multiply enormously and invade nearly all tissues and organs of the tick's body. Within 3 days they begin to arrive in the salivary glands, the coxal organs and ovaries. In *O. moubata* the salivary glands in the nymphs appear to be more heavily infected than do those of the adults, in which the infection tends to diminish and die out. In contrast the coxal organs of the nymphs are usually only lightly infected whereas those of the adults become heavily infected. When either the immature stages or adults of *O. moubata* feed on humans, or some other host, saliva is injected inter-mittently into the bite and spirochaetes can be introduced by this route, especially by nymphal ticks. During feeding excess body fluids are filtered from the haemocoel by the coxal organs and in infected ticks, especially adults, the coxal fluids contain spirochaetes ingested with a previous blood-meal. These can enter the host through the puncture of the tick's bite or through the intact skin. Humans can therefore become infected with *B. duttoni* by either the bite of *O. moubata* or the coxal fluids, or both.

In other *Ornithodoros* species changes in the level of spirochaete infec-tion with age in the salivary glands and coxal organs do not necessarily follow the same pattern. Moreover, other *Ornithodoros* species tend to excrete excess fluids only when they have left their host, and hence trans-mission by these species is mainly by the bites of the ticks. In no species of *Ornithodoros* is infection spread by faeces.

The tick itself is usually regarded as the most important reservoir, espe-cially as there is hereditary (trans-ovarial) transmission, that is the ovaries of adult female ticks become infected with spirochaetes which are then passed on to the eggs so that the newly hatched larvae, and all nymphal instars and adults of both sexes, are infected. Thus, although the larvae, nymphs and adults may not have fed on a host infected with *B. duttoni* they can nevertheless be infected and transmit the disease to other hosts. This phenomenon is called trans-ovarial transmission and can be continued for some three to four generations.

Another rather similar and associated method of transmission is trans-stadial transmission. This involves one of the immature stages of a tick becoming infected by biting a host and then passing the infection on to one or more later stages. For example, a larva might become infected by feeding on an infected host and pass the spirochaetes to the nymphs and adults, or the infection might start with a nymph and be passed to subse-quent nymphal instars and the adults. In all cases trans-ovarial transmis-sion can follow.

16.4.2 Q fever
Q fever is a rickettsial disease caused by *Coxiella burneti*. It has a more or less world-wide distribution and is primarily an infection of rodents and other small mammals and domestic livestock. It can readily be transmitted to people who consume contaminated milk and other foods, and also by the bites of argasid, but mainly ixodid, ticks. Trans-ovarial transmission occurs.

16.4.3 Viruses
More than 100 arboviruses are known to be transmitted by ticks, but only about 30 have been isolated from soft ticks, and very few infect people. Soft

ticks, in marked contrast to hard ticks, are not important medical vectors of arboviruses. (*Ornithodoros* species, however, are the main vectors of African swine fever virus among pigs.)

16.5 CONTROL

The methods used for removing ticks from their hosts are described in Chapter 17 (p. 237).

Suitable repellents that can be applied to the skin include dimethyl-phthalate (dimp), diethyltoluamide (deet), dibutyl phthalate, dimethyl carbamate and indalone. Alternatively, clothing can be impregnated with these chemicals, or with permethrin or cyfluthrin insecticides, to help prevent tick infestations.

Houses infested with argasid ticks, such as *Ornithodoros* species, can be sprayed with insecticides such as 5% HCH, 3% malathion, 0.5% chlorpyrifos (Dursban), 0.5% diazinon, 1% pirimiphos methyl (Actellic), 0.5% bendiocarb (Ficam) or 1% propoxur (Baygon). Special care should be taken to spray floors and cracks and crevices in walls and furniture, and other sites where ticks may be resting. In better houses where cracks are plastered over this usually reduces the numbers of ticks resting in them. In areas where houses have been sprayed with residual insecticides for malaria control there has often been a reduction in the numbers of *Ornithodoros*.

FURTHER READING

Camicas, J.-L., Hervy, J.-P., Adam, F. and Morel, P.-C. (1998) *The Ticks of the World* (Acarida, Ixodida). *Nomenclature, Described Stages, Hosts, Distribution*. Paris: Editions de l'ORSTOM.

Evans, G.O. (1992) *Principles of Acarology*. Wallingford: CAB International.

Griffiths, R.B. and McCosker, P.J. (eds.) (1990) Proceedings of the FAO expert consultation on revision of strategies for the control of ticks and tick-borne diseases, Rome, 25–29 September 1989. *Parassitologia*, **32**, 1–209.

Hoogstraal, H. (1966) Ticks in relation to human diseases caused by viruses. *Annual Review of Entomology*, **11**, 261–308.

Hoogstraal, H. (1967) Ticks in relation to human diseases caused by *Rickettsia* species. *Annual Review of Entomology*, **12**, 377–420.

Hoogstraal, H. (1978) Tickborne diseases of humans: a history of environmental and epidemiological changes. In *Medical Entomology Centenary Symposium Proceedings*, ed. S. Willmott, pp. 48–55. London: Royal Society of Tropical Medicine and Hygiene.

Hoogstraal, H. (1981) Changing patterns of tickborne diseases in modern society. *Annual Review of Entomology*, **26**, 75–99.

Krantz, G.W. (1978) *A Manual of Acarology*, 2nd edn. Corvallis: Oregon State University.

McDaniel, B. (1979) *How to Know the Mites and Ticks*. Dubuque, Iowa: W. C. Brown.

Obenchain, F.D. and Galun, R. (1982) *Physiology of Ticks*. Oxford: Pergamon Press.

Oliver, J.H. (1989) Biology and systematics of ticks (Acari: Ixodoidea). *Annual Review of Ecology and Systematics*, **20**, 397–430.

Sauer, J.R. and Hair, J.A. (1986) *Morphology, Physiology, and Behavioral Ecology of Ticks*. Chichester: Ellis Horwood and New York: Wiley.

Schuster, R. and Murphy, P.W. (eds.) (1991) *The Acari: Reproduction, Development and Life History Strategies*. London: Chapman & Hall.

Sonenshine, D.E. (1991) *Biology of Ticks*, vol. 1. New York and Oxford: Oxford University Press.

Sonenshine, D.E. (1993) *Biology of Ticks*, vol. 2. New York and Oxford: Oxford University Press.

Wilde, J.F.K. (ed.) (1978) *Tick-borne Diseases and Their Vectors*. Edinburgh: Centre for Tropical Veterinary Medicine, University of Edinburgh.

See also references to hard ticks at the end of Chapter 17.

17

Hard ticks (Ixodidae)

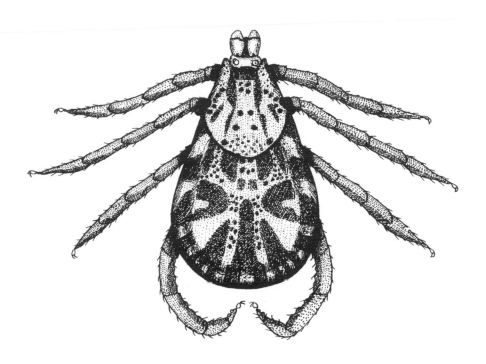

Hard ticks (Ixodidae) have a world-wide distribution, but they occur more frequently in temperate regions than soft ticks (Argasidae). There are about 690 species of hard ticks belonging to 19 genera. From the medical point of view the more important genera are *Ixodes, Dermacentor, Amblyomma, Haemaphysalis, Rhipicephalus* and *Hyalomma*. Some hard ticks are vectors of typhuses such as Rocky Mountain spotted fever (*Rickettsia rickettsi*) and Boutonneuse fever (*R. conori*). In addition some can spread Q fever (*Coxiella burneti*) and many arboviruses – including Russian spring–summer encephalitis, tick-borne encephalitis, Omsk haemorrhagic fever, Kyasanur Forest disease, Colorado tick fever, and Crimean–Congo haemorrhagic fever. They also transmit tularaemia (*Francisella tularensis*), and cause tick paralysis.

17.1   EXTERNAL MORPHOLOGY

Adult hard ticks are flattened dorsoventrally and are oval in shape, measuring about 1–23 mm in length depending on species and whether they are unfed or fully engorged with blood. The females are nearly always bigger than the males, and because they take larger blood-meals they enlarge much more than males during feeding.

The capitulum or 'false head' projects forwards from the body and is visible from above (Fig. 17.1), thus distinguishing adult hard (ixodid) ticks from soft (argasid) ticks (Fig. 16.1). There are also differences in the shape and structure of the capitulum which serve to separate the two families. For example, in hard ticks the palps are swollen and club-shaped (Fig. 17.2), rather than leg-shaped as in the soft ticks. The toothed hypostome is located between a pair of chelicerae which, unlike those of soft ticks, have their cheliceral sheaths covered with very small denticles, giving them what is sometimes termed a 'shagreened' appearance. As in argasid ticks both the hypostome and chelicerae penetrate the host during feeding. The length and shape of the palps and basal part of the capitulum, termed the basis capituli, serve to separate the important genera of ixodid ticks.

The posterior margin of the body in species of the genera *Dermacentor, Rhipicephalus* and *Haemaphysalis* has a number of rectangular indentations called 'festoons'. However, in fully engorged females these indentations may be difficult to see due to the body's distension with blood.

All hard ticks have a dorsal plate called a shield or scutum, which is absent in soft ticks. In males the scutum is large and covers almost the entire dorsal surface of the body (Fig. 17.1), whereas in females it is much smaller and is restricted to the anterior part of the body just behind the capitulum (Fig. 17.1). In both sexes of *Dermacentor, Amblyomma* and some *Rhipicephalus* species the scutum has so-called enamelled coloured areas, and such ticks are described as being ornate species. In engorged females the scutum may be difficult to see because it appears small in relation to the

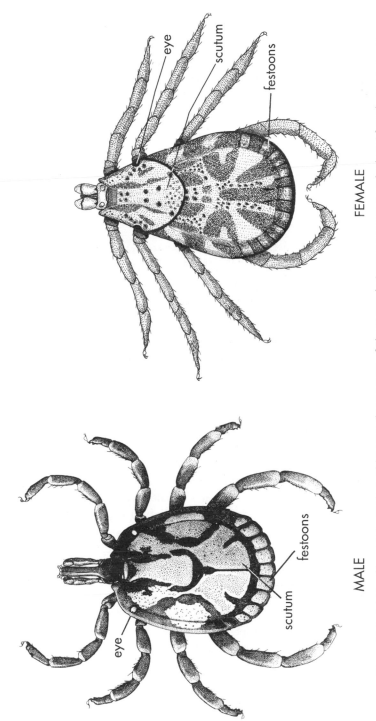

Figure 17.1 Adults of hard ticks. Male *Amblyomma* (by courtesy of The Natural History Museum, London) and female *Dermacentor* showing sexual differences in size of the scutum. Note the presence of festoons.

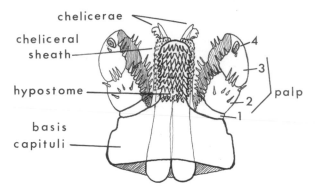

chelicerae

cheliceral
sheath

hypostome

basis
capituli

4
3
2
1
palp

Figure 17.2 Capitulum of an adult ixodid tick showing club-shaped palps with minute fourth segment, and denticulate cheliceral sheaths.

enlarged body and becomes pushed forwards so that it is almost vertical in position. The scutum provides a method of immediately recognizing hard ticks and also of differentiating between the sexes, although medically this may not be very important as both sexes bite animals and are therefore potential disease vectors. In the larval and nymphal stages the scutum is small in both sexes.

The body has four pairs of legs terminating in a pair of claws.

There are no coxal organs in ixodid ticks. Unlike soft ticks, males of some species of hard ticks may have sclerotized plates ventrally, and their number, arrangement and colour may be of taxonomic importance.

17.2 LIFE CYCLE
Both the Ixodidae and Argasidae have hemimetabolous life cycles, that is, there is incomplete metamorphosis involving a larval and nymphal stage. There are, however, important differences between the life cycles and ecology of ticks in these two families. Adult ixodid ticks remain attached to their hosts for long periods as blood-feeding often lasts for 1–4 weeks. When feeding has finished the enormously engorged tick drops from the host to the ground and seeks shelter under leaves, stones, detritus, amongst surface roots of grasses and shrubs, or buries itself in the surface soil. The time taken for females to digest their blood-meal and commence laying eggs varies according to species and environmental conditions, especially temperature. Sometimes oviposition begins 3–6 days after the female drops from the host, but egg laying may not start until several weeks, or occasionally months, after the end of feeding. Thousands (often 1000–8000) of small spherical eggs are laid in a gelatinous mass which is formed in front and on top of the scutum of the tick (Fig. 17.3). Some species lay as many as 20000 eggs and the egg mass may become larger than the ovipositing female. Oviposition may last for 10 days, or extend

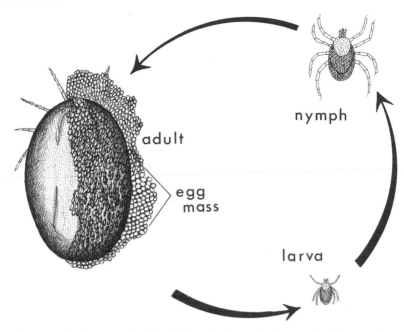

Figure 17.3 Life cycle of an ixodid tick showing a single nymphal stage, and female with very large egg mass.

over about 5 weeks. As in argasid ticks the eggs are coated with a waxy secretion produced by Gene's organ, which in the case of ixodid ticks also helps to transfer the eggs from the genital opening to the scutum. The ixodid female lays only one batch of eggs, after which she dies.

After 10–20 days to several months six-legged larvae hatch from the eggs. The larvae are minute, being about 0.5–1.5 mm long, and are sometimes referred to as 'seed ticks'. On cursory examination they superficially resemble larval mites, but the presence of a *toothed* hypostome immediately identifies them as ticks. After emergence the larvae remain inactive for a few days after which they swarm over the ground and climb up vegetation and cluster at the tips of grasses and leaves. When suitable hosts pass through a tick-infested habitat the larvae respond to stimuli such as carbon dioxide, host odours, warmth, shadows, vibrations and movements by waving their front legs in the air. This host-seeking behaviour exhibited by the larvae, and also the nymphs and adults, is called 'questing'. Larvae climb onto their hosts and crawl to their favoured sites for attachment, commonly in the ears or on the eyelids, but the selected site depends on the species of tick and host. The chelicerae and hypostome are inserted deep into the skin of the host and the larvae commence blood-feeding. The larvae remain attached to their hosts for about 3–7 days, and then drop to the ground and seek shelter amongst vegetation or under stones. Larvae normally take 3–6 days to digest their blood-meals, but in cooler weather digestion may extend over several weeks. After all blood has been digested the larvae remain inactive for a few days before they moult to become nymphs.

The newly formed eight-legged nymphs crawl over the ground and climb vegetation and behave similarly to the larvae (that is questing) in seeking out a suitable passing host. They attach themselves to selected sites on the body and begin to feed, and 5–10 days later the fully engorged nymphs drop to the ground and again seek shelter under stones or amongst vegetation. They remain quiescent for about 3–4 weeks, during which time the blood-meal is digested, and afterwards the nymphs moult to produce male or female ixodid ticks. There is only one nymphal stage in the life cycle of ixodid ticks (Fig. 17.3), whereas argasid ticks have several nymphal stages.

The newly formed adults remain more or less inactive for about 7 days, after which they climb vegetation and start questing for passing hosts. Adult female ticks can take enormous blood-meals. Some ingest up to 600 mg of blood, although much of the fluid of this is excreted; nevertheless the weight of an unfed female can increase some 200 times after blood-feeding. In contrast male ticks ingest very little blood. On the ground adults seek shelter under stones, surface vegetation and amongst roots of plants. Except for *Ixodes* species, mating in ixodid ticks occurs on the host and males may stay attached for several weeks or months, mating with several females.

17.3   BEHAVIOUR AND HABITS

Many species of ixodid ticks are more or less host-specific. For example some species feed almost exclusively on birds, others on reptiles, or on certain types of mammals such as bats, canids or bovids, but some species feed on almost any available hosts including people. Diversity of host species often increases the likelihood of ticks transmitting diseases among their hosts, including humans. Larvae and nymphs of many, but not all, ticks seem to have a predilection for small animals such as rodents, cats and dogs, and ground-inhabiting birds, whereas adults seem to prefer to feed on cattle, horses and a variety of large and wild mammals. Humans are parasitized by all life-stages of ixodid ticks, but often less so by adults than by the younger stages. The tick's life cycle may be prolonged by months or even years by lack of suitable hosts. Some species, for example *Ixodes ricinus*, have a life cycle, from egg to adult, which even under favourable conditions lasts about 3 years. In temperate regions development may also be prolonged or cease temporarily during winter months. In warm countries development and breeding may continue throughout the year, but there may be seasonal fluctuations; unfavourable climate may prevent or delay feeding and development. Adult ticks can withstand starvation for 1–2 years, and may live up to 7 years, but usually they live for only about 2 years.

Although ticks can tolerate considerable variations in temperature and humidity, most species are absent from both very dry and very wet areas, but certain *Hyalomma* species occur in arid areas and deserts. Humidity at soil level can be an important factor in tick survival, and this may be very different from humidities measured at greater heights. Microclimatic conditions at soil level are greatly influenced by the amount and type of ground vegetation, so that the distribution of various species of ticks can often be closely associated with particular types of vegetation.

Both immature and adult ixodid ticks remain on their hosts much longer than do argasid ticks and may be carried many kilometres by their hosts, or even across continents by migrating birds, before they drop off. They are therefore not restricted to their hosts' homes or resting places as are most argasid ticks, but are often more widely dispersed.

17.3.1 Three-host ticks
The life cycle previously described refers to a three-host tick, that is a different individual host, which may be the same or different species, is parasitized by the larva, nymph and adult, and moulting occurs on the ground. Most ixodid ticks have this type of life cycle, and medically important species of three-host ticks are found in the genera *Ixodes*, *Dermacentor*, *Rhipicephalus* and *Haemaphysalis*. Ticks which feed on three hosts are more likely to become infected with pathogens and be potential vectors of disease than species that feed on one or two hosts.

17.3.2 Two-host ticks
Some species of ticks, in particular many of those in the genera *Hyalomma* and *Rhipicephalus*, are two-host ticks. After the larva has completed feeding it remains on the host and moults to produce a nymph which then feeds on the same host. The engorged nymph drops off, moults and the resultant adult feeds on a different host.

17.3.3 One-host ticks
In a few ticks, such as *Boophilus* species, the larva, nymph and adult all feed on the same host and moulting also takes places on that host. The only stage that leaves the host is the blood-engorged female tick which drops to the ground to lay eggs. One-host ticks are less likely to acquire infections with pathogens than ticks which feed on several hosts, and clearly the only method by which infection can be spread from one host to another by these ticks is by trans-ovarial transmission. One-host ticks are of little or no medical importance, but certain species of *Boophilus* are important vectors

of several animal diseases including babesiosis, such as redwater fever (*Babesia bigemina*) which infects cattle, and bovine anaplasmosis (*Anaplasma marginale*).

17.4   MEDICAL IMPORTANCE

17.4.1 Tick paralysis

Females of some hard ticks, especially certain species of *Dermacentor*, *Ixodes* and *Amblyomma*, cause a condition known as tick paralysis. Human cases have been reported from North and South America, Europe, Asia, Australia and South Africa. Ticks in these three genera and also species of *Rhipicephalus* and *Haemaphysalis* can also cause tick paralysis in pets and domesticated animals. The symptoms appear 5–7 days after a tick, usually a female, has commenced feeding. There is an acute ascending paralysis affecting the legs with the result that the person cannot walk or stand, and has difficulty in speaking, swallowing and breathing, due to paralysis of the motor nerves. The symptoms are painless and there is very rarely any rise in the patient's temperature. Tick paralysis can be confused with paralysis due to poliomyelitis and certain other paralytic infections. Young children, especially up to the age of 2 years, are most severely affected. Death in animals, and in rare cases also humans, can result due to respiratory failure. Removal of ticks from patients, usually, but not always, brings about a full recovery after a few days or weeks.

Tick paralysis is not caused by any pathogens but by various toxins contained in the female tick's saliva which is continually pumped into the host during the long period the tick is feeding on the host. Different species of ticks and also different populations of the same species may vary markedly in their ability to produce tick paralysis in humans and animals.

17.4.2 Arboviruses

More than 100 arboviruses have been recovered from ticks, but all the important tick-borne viral diseases of humans are spread by hard ticks. Most are encephalitides, i.e. they produce clinical responses in which encephalitis is the predominant feature; only a few viruses, such as that causing Colorado tick fever, produce generalized infections. All arboviruses are transmitted by the tick's bite, and trans-ovarial transmission usually occurs in several of the tick species involved.

Russian spring–summer encephalitis (RSSE)

RSSE is caused by one of several closely related viruses of the RSSE complex, which vary in their pathogenicity to humans. It is associated with the taiga forests of the former USSR, Siberia, northern Asia and China. The main vector is *Ixodes persulcatus*, but in certain areas *Haemaphysalis concinna* appears to be an important vector. After multiplication in the tick, virus accumulates in the salivary glands, and infection is through the tick's bite. Various small mammals in addition to ticks serve as reservoirs. There is trans-stadial and trans-ovarial transmission.

Tick-borne (central European) encephalitis (TBE)

TBE virus produces a disease with symptoms very similar to that of RSSE. It occurs in central Europe from Scandinavia to the Balkans. The principal vectors are probably *Ixodes ricinus* and *Dermacentor marginatus*, but the virus has also been isolated from *Haemaphysalis* species. Various small forest mammals, such as rodents, insectivores, as well as dogs and foxes, serve as reservoirs of infection and some have an amplifying role. TBE virus accumulates in the mammary glands of goats, sheep and cows, and people may become infected by drinking infected unpasteurized milk or eating cheese. As with RSSE most infections occur in the spring and summer months. Both trans-stadial and trans-ovarial transmission occur.

Omsk haemorrhagic fever (OHF)

The virus causing OHF is antigenically very similar to viruses causing RSSE, TBE and KFD, and the clinical symptoms are rather similar to those caused by these other viruses. OHF occurs in the western Siberian region of the former USSR. The primary tick vectors are species of *Dermacentor* and *Ixodes* which feed on a variety of rodents, especially muskrats which are amplifying hosts. Infections acquired from these animals are transmitted trans-stadially to nymphal or adult stages and then to humans. Trans-ovarial transmission also occurs. Muskrat hunters are particularly liable to come into contact with infected ticks, and the disease can also be passed directly to them by the animals' urine and faeces.

Kyasanur Forest disease (KFD)

KFD was first recognized in 1957 when monkeys were dying in Kyasanur Forest in Karnataka State of southern India and people were also becoming ill and dying. The main vector is *Haemaphysalis spinigera*. Larval ticks feed on birds and small forest mammals, whereas the nymphal stages feed mainly on monkeys and humans. Larger mammals such as deer, bison and cattle brought to the forest edge to graze serve as hosts for adult ticks, which also bite people. Monkeys are regarded as the principal amplifying hosts. There is trans-stadial but apparently no trans-ovarial transmission in *H. spinigera*. KFD has been found in other forests in southern India and is associated with deforestation and people's intrusion into foci of infection. The epidemiology of KFD is particularly interesting because it shows how changes in people's behaviour, deforestation and agricultural development can lead to changing ecology and disease outbreaks in the human population.

Crimean–Congo haemorrhagic fever (CCHF)

CCHF occurs in the former USSR, especially in the Crimean region, in at least seven European countries, including France, Portugal and Bulgaria, in Africa from Egypt to South Africa and from Mauritania to Kenya, in the Middle East, in Pakistan, India and China. The disease is enzootic in savannah, steppe and semidesert areas. Transmission in Europe and parts of Asia is mainly by ticks of the *Hyalomma marginatum* complex and by *Dermacentor marginatus*. In other areas vectors include species of *Amblyomma* and *Rhipicephalus*. Ticks occur on a variety of animals, including sheep and cattle – in which the infection is maintained by *Boophilus* species – and birds, many of which migrate from Europe to Africa or Asia. In fact some 5 billion birds fly annually from Europe to Africa, and about half return. Although birds are not reservoirs of infection they can spread infected ticks around much of the world. Transmission is by tick bite or by crushing infected ticks, or by accidental infection when shearing tick-infested sheep. Ticks are regarded as the main reservoir of infection. There is trans-ovarial transmission in *Haemaphysalis* species, but possibly not in most other vectors.

Colorado tick fever (CTF)

CTF is a viral disease that occurs in the Rocky Mountain states and South Dakota of the western USA and in western Canada. The principal vector is *Dermacentor andersoni*. CTF is a zoonosis and the virus is spread by species of *Dermacentor*, *Haemaphysalis* and the argasid tick *Otobius lagophilus* among rabbits and hares, and other rodents. Squirrels and woodrats are, together with ticks, the main reservoirs of infection. Adult ticks feed on larger mammals, such as deer, cattle and people. There is trans-stadial but apparently no trans-ovarial transmission.

Miscellaneous arboviruses

Ixodid ticks spread many other arboviruses some of which infect humans, such as Powassan encephalitis (POW) virus in North America and the former USSR, Langat (LGT) virus in Malaysia, and Louping Ill (LI), an important disease of sheep in Britain, which is spread by *Ixodes ricinus* and occasionally infects humans.

17.4.3 Rickettsiae

Rocky Mountain spotted fever (RMSF)

RMSF is also known as Mexican spotted fever, São Paulo spotted fever, American tick-borne typhus and by several other local names. Different strains of the causative organism, *Rickettsia rickettsi*, vary considerably in their virulence. Certain non-pathogenic rickettsiae are also transmitted by vectors of RMSF and are often confused with *R. rickettsi*. RMSF occurs in North, Central and South America, where it can cause death in humans not treated with antibiotics. The principal vectors in North America are *Dermacentor andersoni* (western USA) and *D. variabilis* (eastern USA), whereas in South America *Amblyomma cajennense* is an important vector. RMSF is a zoonosis and dogs, rabbits and small rodents also become infected, and the disease is spread amongst them by various ticks belonging to several genera. Despite short-lived viraemias (5–8 days) small rodents such as mice and voles can act as amplifying hosts, but the tick itself is considered the main reservoir of infection, especially as infections can persist in overwintering ticks. Dogs are not considered as either reservoirs or amplifying hosts, but they can transport infected ticks to human habitations where they may become dislodged and attack humans.

There is an incubation period of about 9–12 days before an infected tick becomes infective, and transmission is normally through the bite of any stage in the life cycle of the tick. An infective tick, however, must remain feeding on a host for at least 2 hours before sufficient rickettsiae are injected into the host for the host to become infected. Consequently, early tick removal may prevent transmission. There is both trans-stadial and trans-ovarial transmission, but the latter is more likely to occur if the infection is picked up by larvae and nymphs than by adult females.

Siberian tick typhus (STT) or North Asian tick typhus (NATT)

STT is similar to RMSF, and the causative agent, *Rickettsia sibirica*, is antigenically close to *R. rickettsi*. It occurs in the former USSR, the Czech and Slovak Republics, Pakistan, Afghanistan, Pacific areas and on Japanese islands. Vectors include species of *Dermacentor*, *Haemaphysalis*, *Hyalomma* and *Rhipicephalus*. Infection is by tick bites, and both trans-stadial and trans-ovarial transmission occur. Ticks appear to be the main reservoir of infection. Small mammals, mainly rodents, may also serve as reservoirs but since the infection in these animals is short-lived they are probably not important in maintaining the reservoir.

Boutonneuse fever

Also known as fièvre boutonneuse, Marseilles fever, South African tick typhus, Kenyan tick typhus, Indian tick typhus, Crimean tick typhus, etc., boutonneuse fever is caused by *Rickettsia conori*. The symptoms in people are similar to those caused by RMSF. It occurs in the Mediterranean region, the Middle East, Crimea, most of India, South-east Asia and Africa. In the Mediterranean area the principal vector is *Rhipicephalus sanguineus*, the dog tick, but it appears that ticks of several ixodid genera can transmit the disease to humans. Transmission is by the tick's bite, and both trans-stadial and trans-ovarial transmission occur. Ticks, various rodents and, in contrast to RMSF, dogs can be reservoirs. Infection can also occur if infected ticks are crushed and the rickettsiae rubbed into abrasions or the eyes.

Miscellaneous rickettsiae

A disease known as Queensland tick typhus (*Rickettsia australis*) is transmitted by *Ixodes* ticks and occurs in the Queensland area of Australia. There is both trans-stadial and trans-ovarial transmission.

Another rickettsial disease is Q fever (*Coxiella burneti*), which has a world-wide distribution and is mainly a disease of rodents and other mammals. It is mainly spread by consuming contaminated milk and meat from cattle, and the inhalation of dried infected faeces by those working with cattle, sheep and goats. Hard ticks (and also soft ticks) may help maintain an enzootic cycle, and even spread the infection to humans by their bites or infected faeces. Trans-ovarial transmission occurs.

17.4.4 Ehrlichiosis

Species in the genus *Ehrlichia* cause illness in dogs, horses and deer and more rarely in humans. One species, *E. chaffeensis*, causes human granulocytic ehrlichiosis (HGE), and more recently another species has been found to cause human monocytic ehrlichiosis (HME). Various hard ticks are the vectors. There seems to be trans-stadial transmission and possibly transovarial transmission. Most human cases occur in the USA, where the disease seems to be increasing in prevalence, but ehrlichiosis has also been reported from Europe, Africa and Venezuela.

17.4.5 Spirochaetes

Lyme disease

Lyme disease, which was first recognized in 1975 in the town of Old Lyme, Connecticut, USA, is caused by a spirochaete called *Borrelia burgdorferi*, one of three species within the *B. burgdorferi* complex. In the eastern USA transmission to humans is by the bite of *Ixodes scapularis* (*I. dammini*) whereas in western areas *I. pacificus* is the principal vector. The more normal hosts of these ticks are deer. Lyme disease is a zoonosis. In the USA prevalence is increasing and Lyme disease has become the most important vector-borne disease, with over 16 000 cases reported in 1997. It also occurs in Canada, most of Europe, Asiatic and European parts of the former USSR, China, Japan, and possibly in Australia. In these areas other species of *Ixodes* (mainly *I. ricinus* in Europe and north Africa and *I. persulcatus* in Asia) are the vectors. Transmission is by tick bite and there is both trans-stadial and trans-ovarial transmission. At least 15 species of wild animals (mainly small rodents) as well as cows, horses and dogs have been found to act as reservoirs of infection. Birds can also harbour certain strains of *B. burgdorferi*.

The ecology of Lyme disease and reasons for its increased prevalence and extension of its geographical range are both complex and interesting. It seems that one explanation is that in both North America and Europe people are spending more time in rural areas where there can be large infestations of ticks from deer. Also more people are building homes next to recently cleared land adjacent to forests and this increases their chances of being bitten by ticks. Increased awareness and more extensive serological testing of people may also be partly responsible for the increase in numbers of reported cases.

17.4.6 Tularaemia

Tularaemia is a bacterial disease caused by *Francisella tularensis* (=*Pasteurella tularensis*), which occurs in North America, Europe, Japan and Asia. It infects mainly rabbits, but other rodents and even birds can be infected. The disease is spread by a variety of direct contact methods such as handling infected live animals or carcasses, drinking contaminated water, eating raw and uncooked meats and also by the bites of various hard ticks. *Chrysops discalis* can also be a vector (p. 117).

17.5   CONTROL

Many methods are advocated for removal of ticks from their hosts including coating them with vaseline, medicinal paraffin and nail varnish, all of which block their spiracles and thus prevent ticks from breathing. But it may be several hours before such ticks will withdraw their mouthparts and this is usually unacceptable, especially as rapid removal of ticks often reduces the chances of disease transmission. More rapid removal is sometimes achieved by dabbing ticks with chloroform, ether, ethyl acetate, benzene or some other anaesthetic. Recent research in the USA indicates that the best removal method is the simplest, that is to pull ticks out as soon as possible with blunt forceps and treat the bite wound with an antiseptic agent.

Several species of hard ticks transmit diseases to livestock and in many areas of the world regular dipping of sheep and cattle in acaricidal baths, or spraying them with acaricides (insecticides) is practised. Intensive use of chemical control has resulted in ticks becoming resistant to some organochlorine, organophosphate, carbamate and even synthetic pyrethroid insecticides. Despite this, acaricides are still needed and used to control ticks that may be pests or vectors of human diseases. For example, acaricidal formulations can be applied to domestic pets, such as dogs, to rid them of their ticks. Recommended treatments include solutions of 5% malathion, 0.1% dichlorvos (DDVP), 1% carbaryl (Sevin), 0.1% dioxathion, 0.2%

naled (Dibrom) and 1% coumaphos. Alternatively, dusts of 5% carbaryl or 0.5% coumaphos, 3–5% malathion or 1% trichlorphon can be applied to the coats of pets. Floors of houses, porches, verandahs and other sites where infected pets sleep should be sprayed with oil solutions or emulsions of insecticides as given on p. 223 for control of endophilic soft ticks. Such spraying will kill not only ticks that have dropped off but often those still attached to domestic pets.

Hard ticks outside houses, such as in gardens, yards and nearby fields, can be controlled by ground applications of diazinon, chlorpyrifos (Dursban), carbaryl (Sevin), propoxur (Baygon), permethrin or other synthetic pyrethroids using dosages in the range 1–8 kg/ha. Pellet formulations are best because they more easily penetrate vegetative ground cover and reach the microhabitats of the ticks; a single application may be effective for 6–8 weeks. ULV (ultra-low-volume) spraying with these chemicals reduces dosage rates to about 0.34–1.2 litres/ha. There may, however, be environmental objections to such blanket-type spraying of vegetation.

FURTHER READING

Camicas, J.-L., Hervy, J.-P., Adam, F. and Morel, P.-C. (1998) *The Ticks of the World* (Acarida, Ixodida). *Nomenclature, Described Stages, Hosts, Distribution*. Paris: Editions de l'ORSTOM.

Gothe, R. and Neitz, A.W.H. (1991) Tick paralyses: pathogenesis and etiology. *Advances in Disease Vector Research*, **8**, 177–204.

Lane, R.S., Piesman, J. and Burgdorfer, W. (1991) Lyme borreliosis: relation of its causative agent to its vectors and hosts in North America and Europe. *Annual Review of Entomology*, **36**, 587–609.

Needham, G.R. and Teel, P.D. (1991) Off-host physiological ecology of ixodid ticks. *Annual Review of Entomology*, **36**, 313–52.

Oliver, J.H. (1989) Biology and systematics of ticks (Acari: Ixodoidea). *Annual Review of Ecology and Systematics*, **20**, 397–430.

Reed, G.H. (1993) Lyme disease and other tick-borne diseases. *Journal of Environmental Health*, **55**, 6–10.

See also references given to soft ticks at the end of Chapter 16.

18

Scabies mites (Sarcoptidae)

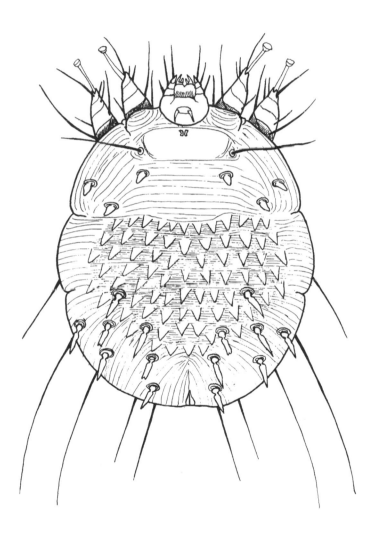

The scabies or itch mite which occurs on people belongs to the species *Sarcoptes scabiei* and has a world-wide distribution. Mites which cannot be separated morphologically from *S. scabiei* are found on numerous wild and domesticated animals, including dogs and pigs, but such mites very rarely infect humans; however, if they do the infection can persist for several weeks or even months. Scabies mites on animals are considered to be the same species as those found on people but physiologically adapted for life on non-human hosts.

Scabies mites are not vectors of any disease but cause conditions known in people as scabies, acariasis, Norwegian itch or crusted scabies. There are at least 300 million scabies cases annually. In animals the condition caused by scabies mites is referred to as mange.

18.1   EXTERNAL MORPHOLOGY

The female mite is just about visible without the aid of a hand lens (0.30–0.45 mm). It is whitish and disc-shaped. Dorsally the mite is covered with numerous small peg-like protuberances and a few bristles, and both dorsally and ventrally there are series of lines across the body giving the mite a striated appearance (Fig. 18.1). Adults have four pairs of short and cylindrical legs divided into five ring-like segments. The first two pairs of legs end in short stalks called pedicels which terminate in thin-walled roundish structures often termed 'suckers'. In the females the posterior

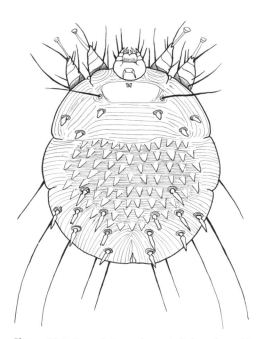

Figure 18.1 Dorsal view of an adult female scabies mite, *Sarcoptes scabiei*.

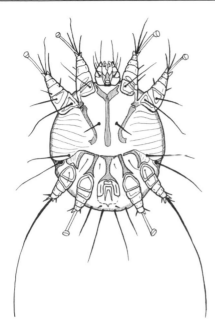

Figure 18.2 Ventral view of the smaller adult male *Sarcoptes scabiei* showing 'suckers' on hind-legs (absent from hind-legs of females).

two pairs of legs do not have 'suckers' but end in long and very conspicuous bristles. There is no real distinct head, but the short and fat palps and pincer-like chelicerae of the mouthparts protrude anteriorly from the body.

Adult male scabies mites are only about 0.20–0.25 mm long, and apart from their small size may also be distinguished from females by the presence of 'suckers' on the last pairs of legs (Fig. 18.2).

18.2   LIFE CYCLE

Scabies was for a long time associated with poverty, overcrowding and poor hygiene, but surveys show that people from all socioeconomic backgrounds can be infested with mites.

The female mite selects places on the body where the skin is thin and wrinkled, such as between the fingers, wrists, elbows, feet, penis, scrotum, buttocks and axillae. The majority (63%) of mites are found on the hands and wrists and about 11% occur on the elbows. The mite digs and eats her way into the surface layers of the skin – the stratum corneum. In women mites may often be found burrowing beneath and around the breasts and nipples. Occasionally mites are found on the face and scalp, especially in the postauricular fold, and on other parts of the body. Often the greatest number of mites on children up to a year old are found on the feet. When the females have burrowed into the superficial layers of the skin they excavate winding tunnels at the rate of about 0.5–5 mm per day, which are seen on the skin as very thin twisting lines a few millimetres to several centimetres long. The mite feeds on liquids oozing from dermal cells she has chewed. She lays 1–3 eggs a day in her tunnel.

The eggs hatch within 3–4 days, and small six-legged larvae emerge which look like miniature adults. These larvae crawl out of the tunnels onto the surface of the skin where a large number die, but a few succeed in either burrowing into the stratum corneum or entering a hair follicle to produce not a tunnel but a small pocket called a 'moulting pocket'. After 3–4 days the larva moults in the pocket to produce an eight-legged nymph (protonymph). After 2–3 days this first nymphal stage moults to produce a second nymph (tritonymph); in mites destined to become females this second nymphal stage is considerably larger than the tritonymph of males. After a further 2–3 days the second nymphal stage (tritonymph) moults to become either a male or female adult. After mating the female adult increases considerably in size and is sometimes referred to as an ovigerous or mature female. Only after fertilization does the female commence to burrow through the skin, and after 3–5 days begins to lay eggs in her tunnel. Female mites rarely leave their tunnels, but contrary to previous beliefs dislodged healthy female mites can sometimes be found on bed linen and on clothing.

Adult male mites are about half the size of ovigerous females and can be found in either very short burrows (usually less than 1 mm) or in small pockets in the skin. However, they probably spend most of their life wandering around on the surface of the skin seeking out female mates.

The life cycle from egg to adult usually takes 10–14 days. Female scabies mites may live for about a month on humans; away from their hosts they may survive for up to 10 days or more if temperatures are relatively low (e.g. 10 °C) and humidities high (e.g. 95% RH), but most probably live for just 2–5 days.

Scabies is a contagious complaint which is transmitted only by close contact. It is therefore a family disease spreading amongst those living in close association, especially when they sleep together in the same bed. It can be spread amongst courting couples who are habitually holding hands. It appears that the actual transfer of mites from person to person takes about 15–20 minutes of close contact. The incidence of scabies often increases during wars and disasters, such as earthquakes, floods and famines, when people are sleeping and living in very overcrowded situations. It is also possible to get infected by sleeping in a bed formerly occupied by an infected person, but this probably rarely happens. As mites may remain alive on discarded clothing for several days it is possible that some may be able to infest a new host, but such indirect transmission seems very rare. It is therefore probably not worthwhile routinely fumigating or sterilizing clothing or bedclothes to prevent scabies spreading, but in epidemics, or in cases of Norwegian or crusted scabies resulting from the use of immunosuppressive drugs, such measures may be needed. Ten minutes at 50 °C will kill the mites; alternatively clothing and bedding can be kept unused for about 4 days, which usually results in their death. Laundering will also kill any mites on clothing.

Increases in the incidence of scabies often appear to go in 15–20 year cycles, probably due to fluctuations in levels of immunity in the human population, although this explanation has been disputed.

18.3 RECOGNITION OF SCABIES

Scabies in humans can be diagnosed by detection of the female mite's thin twisting tunnels; faeces deposited in these tunnels may be visible through the skin as dark pepper-like spots. Unfortunately tunnels are usually not easy to see, especially on dark-skinned people. However, they can often be made visible by smearing a drop of ink on suspected skin; after a few minutes the ink tends to seep into the tunnels and when the excess is wiped off the ink-filled tunnels become visible. Alternatively if liquid tetracycline is used the tunnels fluoresce bright yellow-green under ultra-violet light. The surface layers of skin at the end of the tunnels can be gently scratched away with sharp dissecting needles and the mites, which usually readily adhere to the points of the needles, removed and examined under a ×50 magnification. Alternatively a cruder procedure is to scrape the affected skin areas with a scalpel blade or razor and to examine the scrapings.

The average number of mature adult female scabies mites found on a person is about 11, most patients have 1–15 mites, only about 3% have more than 50 mites.

18.3.1 The scabies rash

The scabies rash is a follicular papular eruption that occurs mainly on areas of the body not infected with burrowing mites, such as the buttocks and around the waist and the shoulders, but the rash can also occur on other parts of the body such as the arms, calves and ankles. It does not appear on the head, centre of the chest or back, nor on the palms of the hands or soles of the feet. The rash is produced in response to the patient being sensitized, that is the rash is an allergic reaction produced by the mites. Frequently patients are unaware they are harbouring mites until a rash appears.

When a person is infected for the first time with itch mites the rash does not usually appear for 4–8 weeks, although in exceptional cases the rash may occur within 2 weeks. In individuals who have previously been infected a rash may develop within a few days after reinfection. The rash may persist for several weeks after all scabies mites have been destroyed. The severe pruritus which soon develops results in vigorous and constant scratching, especially at night, and this frequently leads to the develop-ment of secondary bacterial infections. These may be quite severe leading to boils, pustules, ecthyma, eczema and impetigo contagiosa. These com-plications tend to mask the nature of the complaint, and as a consequence correct diagnosis of scabies may not be made. The seriousness of the symp-toms is not always directly related to the number of mites, and severe reac-tions may be found on people harbouring a few mites.

The condition in Europe known as Norwegian or crusted scabies is rare but highly contagious due to the vast numbers of mites in exfoliating skin scales. It is characterized by the formation of thick keratotic crusts over the hands and feet, scaling eruptions on other parts of the body, and usually large numbers of mites, but a much less pronounced degree of itching. It is not clear why the condition develops, but it may be due to a loss of immu-nity in humans which allows the establishment of enormous numbers of mites. This hypothesis is supported by the finding that the development of Norwegian scabies has sometimes been associated with the extensive use of corticosteroids, and with HIV patients. In such people irritation is reduced and consequently there is less scratching, an act that probably helps in the removal and destruction of mites.

18.4   TREATMENT OF SCABIES

All cases of scabies can be cured; there are no resistant infections. Methods aimed at killing the mites will do little to immediately alleviate the nui-sance and irritation caused by the rash, although this will eventually disap-pear. Separate medical treatment, however, may be necessary especially if secondary infections have become established. In the past, a common pro-cedure was to give the patient a hot bath and a vigorous scrubbing with a brush until the patient bled, but this is not very effective at either removing or killing the mites. However, as many, but certainly not all, patients with scabies are dirty, an ordinary bath before treatment may be advisable for general hygienic reasons. However, if large numbers of patients suffering from scabies are to be treated, such as in epidemic situations, bathing may not be practical.

A 20–25% benzyl benzoate emulsion can be painted on a patient from the neck downwards, and after allowing some 5–10 minutes for this appli-cation to dry the patient can re-dress. A single efficient treatment should in most cases result in a complete eradication of the mites, but a repeat treat-ment on the third day may be advisable. Only in rare cases does dermatitis result from the use of benzyl benzoate and this is more likely to happen in young children.

Mitigal is a yellowish oily liquid sulphur preparation which is painted undiluted over the body from the neck downwards. A single treatment should be 100% effective. A mild form of dermatitis may be produced in some people. Tetmosol is another sulphur compound sometimes used to treat scabies. It is slow in its action and usually about three treatments 24 hours apart are recommended for a complete cure. It is therefore of limited use in mass treatments. It has been combined with soap and sold as Tetmosol soap, and in this form, when regularly used in washing and bathing, it has a slow curative effect and also acts as a prophylactic.

Although these two sulphur preparations and benzyl benzoate are still used, a better treatment for scabies consists of applying a 1% HCH cream or lotion, 1% malathion aqueous emulsion or 5% permethrin cream (Elimite). Although, as with benzyl benzoate and Mitigal, a single treatment has a high success rate, a second application 2–7 days later, if this proves possible, ensures a complete cure. Crotamiton (Eurax) applied as a 10% cream or lotion is a very safe treatment, but two to five daily applications are needed. More recently it has been shown that ivermectin given as a single oral dose of 100–200 µg/kg body weight is effective in killing scabies mites.

With a highly contagious condition such as scabies it is important to treat all members of a family or community living in close association, not just the individual with a particularly bad infestation of mites, otherwise reinfestation will soon occur.

FURTHER READING

Arlian, L.G. (1989) Biology, host relations, and epidemiology of *Sarcoptes scabiei*. *Annual Review of Entomology*, **34**, 139–61.

Azad, A.F. (1986) Mites of public health importance and their control. WHO/VBC/86.931. Geneva: World Health Organization mimeographed document.

Christophersen, J. (1986) Epidemiology of scabies [letter]. *Parasitology Today*, **2**, 247–8.

Daisley, H., Charles, W. and Suite, M. (1993) Crusted (Norwegian) scabies as a prediagnostic indicator for HTLV-1 infection. *Transactions of the Royal Society of Tropical Medicine and Hygiene*, **87**, 295.

Glaziou, P., Cartel, J.L., Alizieu, P., Briot, C., Moula-Pelat, J.P. and Martin, P.M.V. (1993) Comparison of ivermectin and benzyl benzoate for treatment of scabies. *Tropical Medicine and Parasitology*, **44**, 331–3.

Mellanby, K. (1943) [reprinted 1972 with minor additions]. *Scabies*. Middlesex: E.W. Classey.

Orkin, M. and Maibach, H.T. (eds.) (1985) *Cutaneous Infestations and Insect Bites*. [See Chapters 1–18 on scabies]. New York: Marcel Dekker.

Paller, A.S. (1993) Scabies in infants and small children. *Seminars in Dermatology*, **12**, 3–8.

Robinson, R. (1985) Fight the mite, ditch the itch. *Parasitology Today*, **1**, 140–2.

Stanton, B., Khanam, S., Nazrul, H., Nurani, S. and Khair, T. (1987) Scabies in urban Bangladesh. *Journal of Tropical Medicine and Hygiene*, **90**, 219–26.

Van Neste, D.J.J. (1986) Immunology of scabies. *Parasitology Today*, **2**, 194–6.

19

Scrub typhus mites (Trombiculidae)

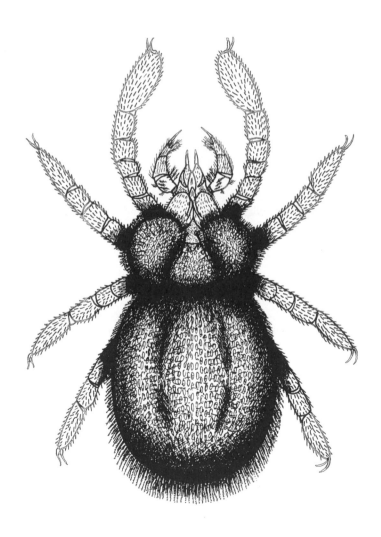

There arc some 3000 species of trombiculid mites belonging to several genera but only about 20 species are medically important because they commonly attack people. Larvae of trombiculids are often called red bugs, chiggers, scrub typhus mites or itch mites. The Trombiculidae have a more or less world-wide distribution, but the medically important species such as the *Leptotrombidium deliense* group of species, including *L. akamushi* and *L. fletcheri*, are found only in Asia. These and other species of *Leptotrombidium* are vectors of scrub typhus *Orientia* (=*Ricksettsia*) *tsutsugamushi* in Asia and the Pacific region including north-east Australia.

Other trombiculid mites in many parts of the world cause itching and a form of dermatitis known as scrub-itch, autumnal itch or trombidiosis. In northern Europe larvae of *Neotrombicula autumnalis* and in North America and parts of Central and South America larvae of *Eutrombicula alfreddugusi* commonly attack people and cause considerable discomfort.

19.1 EXTERNAL MORPHOLOGY

19.1.1 Adults and nymphs
Adults are small mites (1–2 mm), usually reddish and covered dorsally and ventrally with numerous feathered hairs giving them a velvety appearance. There are four pairs of legs ending in paired claws. The body is distinctly constricted between the third and fourth pairs of legs giving it an outline resembling a figure of eight. The palps and mouthparts project in front of the body and are clearly visible from above (Fig. 19.1).

The nymph resembles the adult but is smaller (0.5–1.0 mm) and the body is less densely covered with feathered hairs.

Neither the adults nor the nymphs are of direct medical importance; they do not bite humans or animals but feed on small arthropods and their eggs. It is only the larvae which are parasitic and hence responsible for the spread of diseases.

19.1.2 Larvae
Larvae are very small (0.15–0.3 mm), but after engorging they may increase sixfold in size. They are usually reddish or orange but may be pale yellow or straw-coloured. There are three pairs of legs which terminate in a pair of relatively large claws. Both legs and body are covered with fine feathered hairs. The five-segmented palps and mouthparts are large and conspicuous, giving the larvae the appearance of having a false head (Fig. 19.2). Dorsally and on the anterior part of the body there is a rectangular or pentagonal-shaped scutum bearing three to six setae. However, as the scutum is weakly sclerotized it is often difficult to see under the microscope, unless the light is correctly aligned. More easily detected are a pair of eyes on either side of the scutum. In medically important species there are five

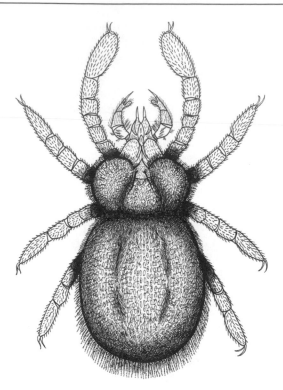

Figure 19.1 Dorsal view of an adult trombiculid mite of the genus *Leptotrombidium.*

feathered setae on the scutum and in addition a pair of specialized feathered hairs known as sensillae which arise from distinct bases. The combination of a body covered with feathered hairs, five scutal hairs, a pair of flagelliform sensillae and large pigmented eyes distinguishes larvae of *Leptotrombidium* from larvae of other mite genera.

19.2 LIFE CYCLE

The life cycle of trombiculid mites is rather complex and unfortunately several different names have been used for the various larval and nymphal stages (see summarized life cycle on p. 250).

Adult trombiculid mites are not parasitic but live in the soil feeding on a variety of small soil-inhabiting arthropods and their eggs. A female mite lays one to five spherical eggs each day on leaf-litter and the surface of damp but well-drained soil, such as river banks, scrub-jungle, grassy fields and neglected gardens. In hot climates egg laying continues uninterrupted for a year or more, but in cooler areas of South-east Asia, including Japan, oviposition apparently ceases during the cooler months of the year, and adults enter into partial or complete hibernation.

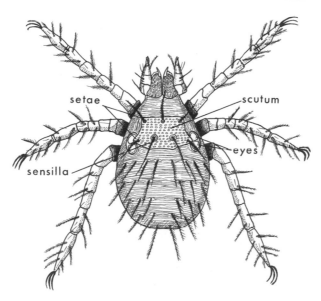

Figure 19.2 Dorsal view of a larval *Leptotrombidium* mite. Note the scutum bearing paired sensillae and five scutal setae.

After about 4–7 days the eggshell splits, but the six-legged larva does not yet emerge but remains within the eggshell and is called the deutovum. After about 5–7 days the larva crawls out of the eggshell and usually becomes very active and swarms over the ground and climbs up grasses and other low-lying vegetation. Larvae attach themselves to birds and mammals especially rodents, and also people, who are walking through infested vegetation. When the larval mites have climbed onto a suitable host they congregate where the skin is soft and moist, such as the ears, genitalia and around the anus. On people larvae seek out areas where clothing is tight against the skin, such as around the waist or the ankles.

Larvae piece the host's skin with their powerful mouthparts and inject saliva into the wound which causes disintegration of the cells. Larvae do not normally suck up blood, rather lymph and other fluid and semidigested materials. The repeated injection of saliva into the wound produces a skin reaction in the host and the formation of a peculiar tube-like structure which extends vertically downwards in the host's skin, which is known as the stylostome or histiosiphon. Some trombiculid mites remain attached to their hosts for as long as 1 month, but the *Leptotrombidium* vectors of scrub typhus remain on people for only about 2–10 days. The engorged larva drops to the ground and buries itself just below the surface of the soil or underneath debris (Fig. 19.3).

The larva having concealed itself becomes inactive and this stage is known as the protonymph. After 7–10 days the protonymph moults to

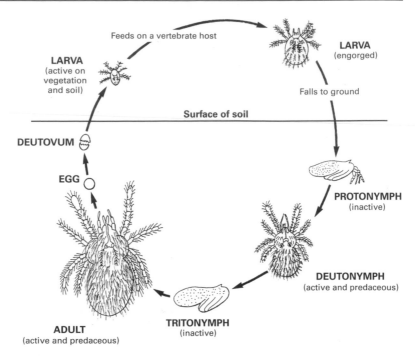

Figure 19.3 Life cycle of a *Leptotrombidium* mite (by courtesy of The Natural History Museum, London).

produce an eight-legged reddish deutonymph covered with feathered hairs. These nymphs are not parasitic, but feed on soil-inhabiting arthropods. After a few days to about 2 weeks the deutonymph ceases feeding and becomes inactive and is called a tritonymph, which after about another 14 days moults to give rise to an adult. The adults resemble the nymphs but are larger, and like them are free-living and feed on small soil-inhabiting animals.

The life cycle (Fig. 19.3) usually takes 40–75 days but may be as long as 8–10 months. The stages in the life cycle can be summarized as follows (the inactive stages are bracketed, and alternative names are in italics):

Egg $\longrightarrow$ (Deutovum, *prelarva*) $\longrightarrow$ Larva $\longrightarrow$ (Protonymph, *first nymphal stage, nymphochrysalis*) $\longrightarrow$ Deutonymph, *second nymphal stage* $\longrightarrow$ (Tritonymph, *third nymphal stage, preadult, imagochrysalis*) $\longrightarrow$ Adult

19.3 ECOLOGY

The free-living nymphs and adults of *Leptotrombidium* have specialized ecological requirements. For example, habitats must contain sufficient numbers of suitable arthropod fauna to serve as food. The habitat must also be one which host animals such as rodents and insectivores regularly traverse so that larvae will have opportunities of attaching themselves to their hosts. Wild rats of the genus *Rattus*, subgenus *Rattus*, are very important hosts of *Leptotrombidium* larvae. In Japan, and possibly elsewhere, small rodents such as species of *Apodemus* and *Microtus* and various subgenera of *Rattus* are also important hosts. Domestic rats play little or no part in the ecology or epidemiology of scrub typhus. In addition to hosts such as these, which maintain the mite population in an area, other more or less incidental hosts, including birds, may be important in aiding the dispersal of larvae to other areas.

Relatively small changes in moisture content of the soil, temperature and humidity can be crucial, as they may cause adults to burrow deeper into the soil and cease egg laying. Habitats favouring the survival and development of the nymphs and adults are therefore in a very delicate ecological balance. Frequently, only very small areas of ground, often just a few square metres, prove to be suitable habitats. This can result in a very patchy distribution of *Leptotrombidium* mites over small areas, but in some situations habitats may comprise several square kilometres. Areas suitable for mite survival and development are often called 'mite islands', a term which emphasizes their frequent isolation from other ecologically suitable areas.

19.4 MEDICAL IMPORTANCE

19.4.1 Nuisance
Several species of Trombiculidae attack humans in temperate and tropical regions of the world. In northern Europe the main pest is *Neotrombicula autumnalis* and in the USA *Eutrombicula alfreddugesi*. Although such mites are not responsible for transmitting any disease, they can nevertheless cause intense itching and irritation, commonly referred to as 'harvest-bug itch', 'autumnal itch' or 'scrub itch'. Larval mites commonly attack the legs. If they are forcibly removed, their mouthparts frequently remain embedded in the skin and this may promote further irritation. People usually become infested with these mites after walking through long grass or scrub vegetation.

19.4.2 Scrub typhus
The causative organism of scrub typhus is *Orientia tsutsugamushi* and the disease is commonly known as scrub typhus, mite-borne typhus, Japanese river fever, chigger-borne rickettsiosis or tsutsugamushi disease. The disease is restricted to Asia and extends from the Primorye regions of the former USSR, Pakistan through India to Myanmar, Indonesia, Malaysia, Thailand, South-east Asia, China, Taiwan, Philippines, Japan, Papua New

Guinea, north-east Australia and neighbouring South-west Pacific islands south to about the tropic of Capricorn. Although most cases are reported from low-lying areas infections occur up to a height of 1000 m in many areas, and even up to about 2000 m in Taiwan and 3200 m in the Himalayas. In India, *Leptotrombidium* mites have been found at elevations of 2700 m. During World War II (1939–45) the incidence of scrub typhus in troops in the Asiatic–Pacific areas was second only to malaria. Over most of the distribution of the disease the principal vector is *L. deliense*, but other species in the *deliense* group, such as *L. akamushi*, are the chief vectors in Japan, whereas *L. fletcheri* is the main vector in Malaysia.

Scrub typhus has often been regarded as a zoonosis, but although *O. tsutsugamushi* can infect forest and rural rodents and *Tupaia* species (tree-shrews) it appears that these animals have a minor role, if any, in the maintenance of scrub typhus. Experimentally it has been shown that it is often difficult to infect trombiculid larvae by feeding them on infected rodents, and even if they do become infected the rickettsiae may not be trans-ovarially transmitted to their progeny. It appears that larvae infected by feeding on humans can pass the infection to their own progeny by trans-ovarial transmission, and subsequently to other humans and rodents. *Leptotrombidium* mites themselves are the main reservoirs of infection.

People become infected following the bite of infected larval trombiculid mites, especially *Leptotrombidium akamushi* and *L. deliense*. People usually get bitten when they visit or work in areas having so-called mite islands, that is patches of vegetation harbouring large numbers of host-seeking larvae. These mite islands may be at the edge of the forest or bush, or on cleared and cultivated land which harbours rodents and is also suitable for the survival and development of the mites. The disease is often associated with 'fringe habitats', that is habitats separating two major vegetation zones such as forests and plantations, because these areas are often heavily populated with rodent hosts. The risk of scrub typhus transmission is often related to the number of areas of different types of vegetation, that is habitat diversity.

All known foci of scrub typhus are characterized by natural or man-made changes in environmental conditions. The very close association between: (a) *Leptotrombidium* mites, (b) wild rodents such as *Rattus* (*Rattus*), (c) transitional secondary vegetation, for example, grass, shrubs and saplings, and (d) *Orientia tsutsugamushi*, has been described as a 'zoonotic tetrad of chigger-borne rickettsiosis'.

Because larval mites attach themselves to only a single host during their life cycle the disease cannot be spread by larvae feeding on one infected host (e.g. humans) and then another. The infection acquired by mites feeding on hosts with rickettsiae is passed on to the free-living nymphal stages and then to the free-living adults. When the female lays her eggs they are infected with rickettsiae and this infection is passed onto the emerging larvae. So, although they have not previously fed on humans they are already infected and consequently transmit the disease to their hosts (humans or rodents) when they feed for the first and only time. This inherited type of transmission is called trans-ovarial transmission and can be maintained for several mite generations before the rickettsiae are reduced in numbers and finally disappear.

19.5 CONTROL

The application to the body of suitable insect repellents such as dimethyl-phthalate (dimp), diethyltoluamide (deet), dibutyl phthalate, dimethyl carbamate or benzyl benzoate may help in reducing the likelihood of people getting infected with mites. Clothing, especially socks, can be impregnated with suitable repellents or with permethrin.

The wide and patchy distribution of chigger mites make their control very difficult, but if mite islands can be identified then it may be possible to remove the scrub vegetation mechanically or by herbicides, and so ensure that the habitat is no longer suitable for the survival of the mites. This, however, frequently is not possible, especially if mites are inhabiting culti-vated land where ground vegetation consists mainly of crops. Spraying areas known or suspected of harbouring mites with residual insecticides such as HCH, dieldrin, fenthion (Baytex), malathion, propoxur (Baygon), toxaphene (camphechlor), diazinon or permethrin, preferably as fogs or emulsions, can do much to reduce the mite population. Insecticidal sprays can be applied from knapsack sprayers, or from equipment mounted on vehicles or aircraft which generates insecticidal aerosols or fogs. Ultra-low-volume (ULV) spraying with approved insecticides such as permethrin or propoxur (Baygon) can also be undertaken.

FURTHER READING

Audy, J.R. (1968) *Red Mites and Typhus*. London: The Athlone Press.

Azad, A.F. (1986) Mites of public health importance and their control. WHO/VBC/86.931. Geneva: World Health Organization mimeographed document.

Buescher, M.D., Rutledge, L.C., Wirtz, R.A., Nelson, J.H. and Inase, J.L. (1984) Repellent tests against *Leptotrombidium (Leptotrombidium) fletcheri* (Acari: Trombiculidae). *Journal of Medical Entomology*, **21**, 278–82.

Krantz, G.W. (1978) *A Manual of Acarology*. 2nd edn. Corvallis: Oregon State University.

McDaniel, B. (1979) *How to Know the Mites and Ticks*. Dubuque, Iowa: W.C. Brown.

Mount, G.A., Grothaus, R.H., Baldwin, K.F. and Haskings, J.R. (1975) ULV sprays of propoxur for control of *Trombicula alfreddugesi*. *Journal of Economic Entomology*, **68**, 761–2.

Roberts, S.H. and Zimmerman, J.H. (1980) Chigger mites: efficacy of control with two pyrethroids. *Journal of Economic Entomology*, **73**, 811–12.

Sasa, M. (1961) Biology of chiggers. *Annual Review of Entomology*, **6**, 221–44.

Traub, R. and Wisseman, C.L. (1968) Ecological considerations in scrub typhus. 1. Emerging concepts. *Bulletin of the World Health Organization*, **39**, 209–18.

Traub, R. and Wisseman, C.L. (1968) Ecological considerations in scrub typhus. 2. Vector species. *Bulletin of the World Health Organization*, **39**, 219–30.

Traub, R. and Wisseman, C.L. (1968) Ecological considerations in scrub typhus. 3. Methods of area control. *Bulletin of the World Health Organization*, **39**, 231–7.

Traub, R. and Wisseman, C.L. (1974) The ecology of chigger-borne ricksettsiosis (scrub-typhus). *Journal of Medical Entomology*, **11**, 237–303.

20

Miscellaneous mites

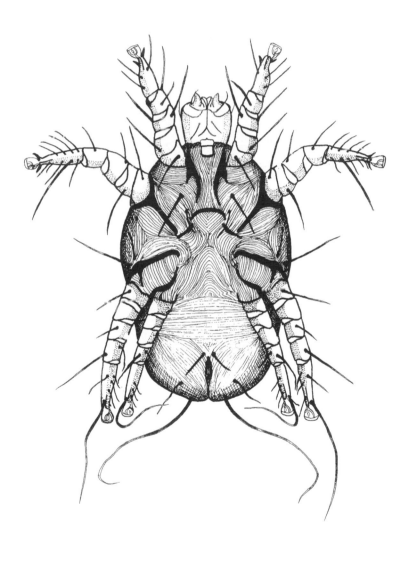

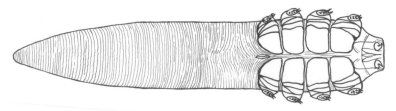

Figure 20.1 Dorsal view of *Demodex folliculorum*, the hair follicle mite, showing four pairs of stumpy legs.

20.1   DEMODICIDAE

20.1.1 *Demodex folliculorum*

Two species of *Demodex* infect humans, namely *Demodex folliculorum* (Fig. 20.1) and *D. brevis*. The former is the more elongate species (0.2–0.4 mm long) and inhabits the hair follicles, whereas *D. brevis* is a squatter species (0.1–0.2 mm) living in the sebaceous glands. Both species have a striated body and four pairs of very short stubby legs and are remarkably non-mite-like.

Demodex mites are therefore found in hair follicles or sebaceous glands, where they feed on subcutaneous tissues, especially sebum. They are particularly common on the nose, eyelids and cheeks adjacent to the nose. They have also been found in ear wax and in the extruded contents of comedones ('blackheads'). Females lay eggs within the hair follicles and these hatch to produce six-legged larvae which moult to give rise to nymphs and finally adults. All the developmental stages, which extend over 13–15 days, occur within the hair follicles or sebaceous glands. Little is known about their biology but apparently a high proportion of adults, especially women, unknowingly have these mites. They are most common in older people, rarely being found in children or adolescents.

Normally they do not appear to produce any adverse effects, although possibly they may sometimes cause dermatitis, such as acne, rosacea, impetigo contagiosa or blepharitis. Daily washing with soap and water can reduce infections. In severe infections resulting in dermatitis 'Danish ointment', which contains a compound polysulphide, can be used, but this should not be applied to the eyelids otherwise irritation may occur. Alternative treatments consist of applying 0.5% selenium sulphide cream, 10% sulphur or 0.5% HCH cream ('Kwell').

20.2   PYROGLYPHIDAE

The family Pyroglyphidae contains five genera, but mites in the genus *Dermatophagoides* are the most important medically. *Dermatophagoides pteronyssinus* is known as the European house-dust mite and *D. farinae* as the

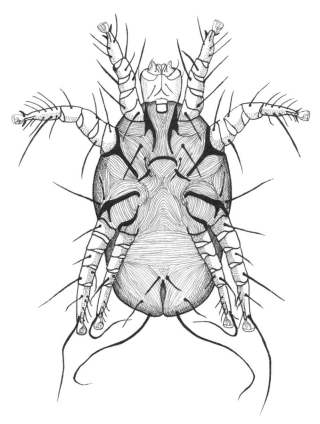

Figure 20.2 Ventral view of a house dust mite, *Dermatophagoides pteronyssinus*.

American house-dust mite, but both species are found more or less world-wide. *Dermatophagoides* (Fig. 20.2) are very small mites (0.3 mm) that live among bed clothes, mattresses, carpets and general house dust. Female mites lay about three eggs a day and these hatch after 6–12 days and a six-legged larva emerges, which feeds and passes through two nymphal stages before becoming an adult; the complete life cycle takes about 3–4 weeks. Beds are the most important, and sometimes only, breeding site. The mites feed on discarded skin scales, semen, scurf and other organic debris, and adults live about 2 months. They are very rarely seen but become airborne when beds are made and are easily inhaled, together with their faeces, and this can cause allergic symptoms resulting in asthma and rhinitis.

Typically there may be 30 mites per gram of house dust, but in Switzerland as many as an estimated 10000 mites per gram of dust have been reported, and in England after bedmaking 44–136 mites per gram in airborne dust have been recorded. Densities above 100 mites per gram are considered a risk factor for sensitization to allergies such as rhinitis, eczema (atopic dermatitis) and asthma, while 500 mites per gram is a major risk factor in the development of acute asthma in those allergic to house-dust mites.

Those suffering from severe allergies caused by house-dust mites can have a series of injections with a specific desensitizing vaccine. Vacuum cleaning carpets may have little effect on removing live mites because they cling firmly to carpet fibres, but may be more effective in removing mites from mattresses and sheets. Enclosing mattresses and pillows in plastic covers can help reduce mite infestations, although sleeping under such conditions may be uncomfortable. Washing bedding above 55 °C or dry cleaning will kill mites, as will leaving electric blankets switched on at maximum temperature for 6 hours or more during the day. Treating sheets with 5% benzyl benzoate can be effective in killing mites, as can spraying carpets with pirimiphos methyl (Actellic) or permethrin. However, it may be difficult to reduce mite densities sufficiently to give any reduction in allergies suffered by householders.

20.3   OTHER MITES

There are numerous other mites which are parasitic on mammals and birds, including pets and livestock, and even on insects, and some of them occasionally become parasitic on people. For example, *Pyemotes tritici* (=*P. ventricosus*) usually parasitizes larvae of grain moths and beetles and is known as the grain mite or hay itch mite, but can bite people handling infested grain and straw. The cosmotropical rat mite (*Ornithonyssus bacoti*), the tropical fowl mite (*O. bursa*) and the chicken mite (*Dermanyssus gallinae*) may sometimes infect people, especially those working closely with infested animals. The bites from these mites can cause irritation and dermatitis. In addition, a few species such as *Liponyssoides sanguineus* which normally is an ectoparasite of mice and rats can spread rickettsial pox (*Rickettsia akari*), and this and a few other species can also transmit to people Q fever (*Coxiella burneti*).

There are also other mites, e.g. so-called forage mites that live among stored products such as grain, copra, flour, dried fruits, cheese and animal feeds. People habitually handling these substances may develop allergic symptoms such as dermatitis, and more rarely bronchitis and asthma. This may lead to terms such as 'grocer's itch', 'baker's itch', and 'copra itch' to describe these occupational hazards.

FURTHER READING

Azad, A.F. (1986) Mites of public health importance and their control. WHO/VBC/86.931. Geneva: World Health Organization mimeographed document.

Cameron, M.M. (1997) Can house dust mite-triggered atopic dermatitis be alleviated using acaricides? *British Journal of Dermatology*, **137**, 1–8.

Cameron, M.M. (1998) Prevention of atopic asthma. *Positive Health*, February, 23–6.

Colloff, M.J., Ayres, J., Carswell, F., Lowarths, P.H., Merrett, T.G., Mitchell, E.G., Walshaw, M.J., Warner, J.O., Warner, J.A. and Woodcock, A.A. (1992) The control of allergens of dust mites and domestic pets: a position paper. *Clinical and Experimental Allergy*, **22**, 1–28.

Fain, A., Guérin, B. and Hart, B.J. (1990) *Mites and Allergic Disease*. Varennes en Argonne, France: Allerbio.

Harvey, P. and May, R. (1990) Matrimony, mattresses and mites. *New Scientist*, 3 March, 48–9.

Krantz, G.W. (1978) *A Manual of Acarology*, 2nd edn. Corvallis: Oregon State University.

McDaniel, B. (1979) *How to Know the Mites and Ticks*. Dubuque, Iowa: W.C. Brown.

Mumcuoglu, Y. (1976) House dust mites in Switzerland. I. Distribution and taxonomy. *Journal of Medical Entomology*, **13**, 361–73.

Nutting, W.B. (ed.) (1984) *Mammalian Diseases and Arachnids*: vol. 1, *Pathogen Biology and Clinical Management*, vol. 2, *Medico-Veterinary, Laboratory, and Wildlife Diseases, and Control*. Boca Raton, Florida: CRC Press.

Owen, S., Morganstern, M., Hepworth, J. and Woodcock, A. (1990) Control of house dust mite antigen in bedding. *Lancet*, **335**, 396–7.

Wharton, G.W. (1976) House dust mites. *Journal of Medical Entomology*, **12**, 577–621.

Glossary of common terms relevant to medical entomology

This glossary contains terms that are used in this book and a few others that are not, but which are considered pertinent to vector biology and control. In general morphological names are excluded as these are adequately explained in the preceding chapters. As usual with glossaries the inclusion and exclusion of terms is somewhat subjective.

Acaricide Chemical that kills mites and ticks. Most acaricides are also insecticides.

Aestivation Condition in which an organism has a period of inactivity as an adaptive response to unfavourable conditions encountered during the hot or dry season, e.g. aestivating adults of some *Anopheles* species.

Afrotropical region (formerly Ethiopian zoogeographical region) The countries of Africa south of mid-Sahara, including southern Arabia, Madagascar, Seychelles and Cape Verde islands. *See* Zoogeographical regions.

Amastigote Morphological form of species of *Leishmania* and *Trypanosoma* with a rounded body and without a flagellum that occurs predominantly in macrophages (*Leishmania* species) or muscle cells (*T. cruzi*) of a vertebrate host.

Amplifying host Usually used in relation to arboviruses where a host attains a very high level of parasites in its blood (high viraemic titre) which makes it very infective to vectors. For example, pigs develop high viraemias of Japanese encephalitis and are important amplifying hosts, as are monkeys in the transmission of Kyasanur Forest disease. Often not an efficient host for long-term maintenance of parasite populations.

Anautogenous Refers to females of blood-sucking insects that require at least one blood-meal to develop their eggs. *See* Autogenous.

Anthropophagic (anthropophilic) Refers to females of blood-sucking arthropods that prefer to feed on humans rather than other hosts. The degree of anthropophagy can vary geographically within species as well as between species. *See* Host preference.

Aperiodic (non-periodic) Not exhibiting periodicity. When applied to microfilariae of helminths it means they are neither periodic nor subperiodic in their appearance in peripheral vertebrate blood, e.g. *Onchocerca volvulus*.

Arbovirus From *ar*thropod-*borne virus*. A virus that multiplies in a blood-sucking arthropod and is principally transmitted by the bite of arthropods to vertebrate hosts, e.g. yellow fever virus. Viruses, such as myxoma virus in rabbits, transmitted mechanically by arthropods (e.g. fleas and mosquitoes) and involving no cyclical development in the vector are not arboviruses.

Australasian region Australia, Tasmania, New Zealand, Melanesia, Micronesia and Polynesia. *See* Zoogeographical regions.

Autogenous Refers to females of blood-sucking insects (usually mosquitoes and simuliids) that lay at least the first batch of eggs without a blood-meal. Subsequently a blood-meal is required for each further oviposition. Some mosquitoes, e.g. *Toxorhynchites* species, never blood-feed and are thus always autogenous. *See* Anautogenous.

Biodegradable insecticide Insecticide that does not accumulate in the environment but is broken down by micro-organisms in the soil and water into relatively harmless compounds, e.g. the organophosphates and carbamates.

Biological control (biocontrol) Deliberate introduction, or augmentation, of biological agents such as pathogens, parasites and predators (especially fish) to control arthropod populations, mainly mosquito larvae. Although *Bacillus thuringiensis* subsp. *israelensis* is often included in this category it is not a biological control agent because it does not recycle and persist in the environment but is applied as a non-living formulation and is thus more accurately described as a microbial insecticide.

Biological transmission Transmission of disease organisms with biological involvement between the vector and parasite. Can involve (i) multiplication without change in form, e.g. *Yersinia pestis* in fleas, arboviruses in ticks and mosquitoes; (ii) multiplication with change in form, e.g. *Plasmodium* species in *Anopheles*; or (iii) no multiplication (usually a decrease in parasite numbers) but with change in form, e.g. filarial parasites in mosquitoes and blackflies.

Borrow pit Excavation, often by the side of a road or railway, dug to provide earth for buildings or embankments. These fill with water and become mosquito larval habitats.

Bromeliad An epiphytic plant growing on trees in the Americas. Its water-filled leaf axils are colonized by mosquito larvae. *See* Epiphyte.

Campestral In epidemiology used to describe transmission occurring in fields and open spaces, such as plague transmission among wild rodents. But also used sometimes to include transmission in woods and forests, which strictly is sylvatic transmission.

Carbamates Synthetic insecticides which are derivatives of carbamic acid, e.g. carbaryl (Sevin) and propoxur (Baygon).

Chemosterilant Chemical used to induce sterility, but not usually death, in arthropods so as to control them, e.g. apholate and tepa. Chemosterilized insects are sometimes used in genetic control of insect vectors.

Chitin A constituent of arthropod cuticle having a hard or leathery texture. (Some insect growth regulators (IGRs) prevent chitin formation and are used in the control of vectors.)

Commensal Animals living together or in close association, e.g. roof rats (*Rattus rattus*) living in and around human habitations.

Cytoform (cytotype) A cytologically defined population with a distinctive chromosomal complement, that may be a species or considered a chromosomal variant within a species. In the latter case individuals are often given vernacular geographical names, e.g. *Anopheles gambiae* Savanna and Forest forms. Different cytoforms may exhibit differences in behaviour.

Dead-end host Host that although infected with an arbovirus, or other pathogen, and may be severely affected, has a viraemia too low in titre to infect blood-feeding arthropods. Examples are humans in the epidemiology of Japanese encephalitis, and horses in western equine encephalomyelitis transmission.

Definitive host Host in which parasites reach maturity. This rarely occurs in arthropod vectors, but the noted exception is the development of malarial parasites, involving sexual reproduction, in mosquitoes. *See* Intermediate host.

Diapause State of inactivity or arrested development which allows an organism to outlast a period of unfavourable conditions. Development cannot be resumed, even under favourable conditions, until diapause is 'broken' by environmental stimuli such as changes in photoperiod (length of daylight) or temperature.

Dichoptic A condition of the head in adult Diptera in which the eyes are widely separated from each other, as, for example, in female adult simuliids. *See* Holoptic.

Diel periodicity Periodicity that occurs about every 24 hours.

Diurnal Refers to activity during the daylight hours, such as blood-feeding in simuliids and appearance in peripheral vertebrate blood of microfilariae of some helminth species, e.g. *Loa loa* in *Chrysops* species.

Emulsion Suspension of miniscule droplets of one liquid in another. For example, when an insecticide dissolved in oil is mixed with water this results in a milky emulsion.

Endemic Describes a disease in a human population that is constantly present and more or less stable. The numbers of new cases cannot exceed the numbers of new susceptible individuals entering the population, but may be fewer. The opposite is epidemic. In animal populations the equivalents are enzootic and epizootic.

Endophagic Describes insects, such as some mosquitoes, that enter houses to blood-feed.

Endophilic Describes insects, such as some mosquitoes, that rest in houses before or after blood-feeding in houses or outside.

Endotoxin Toxin formed inside bodies in Gram-negative bacteria which is released only when the bacteria are lysed (broken down). An example is the toxins in the parasporal body of *Bacillus thuringiensis* subsp. *israelensis*, which on ingestion are lethal to simuliid and mosquito larvae.

Epidemic Occurrence of a disease in the human population where the numbers of cases exceed the normal expected number of cases. An epidemic situation can be only temporary. *See* Endemic.

Epimastigote (crithidial form) Morphological form of a trypanosome with the flagellum emerging about halfway in the body but remaining attached to the cell membrane, e.g. trypanosomes in the salivary glands of tsetse-flies and mid-gut of triatomine bugs. This is not the stage that is infective to the vertebrate host. *See* Metacyclic trypanosome.

Epiphyte Plant that grows on other plants, usually trees, but without being parasitic. It derives its water mainly from rain. Bromeliads, the water-filled leaf axils of which provide mosquito larval habitats, are epiphytes.

Exophagic Term applied to insects that blood-feed outside houses, e.g. *Aedes aegypti* and simuliid species.

Exophilic Term applied to blood-sucking insects that rest outside houses, irrespective of whether they have fed inside or outside houses.

Exoskeleton Outer body layer, often called the integument or cuticle, of arthropods that is shed during moulting from one life-stage to another, e.g. third-instar mosquito larva to fourth-instar, and then fourth-instar larva to pupa.

Extrinsic incubation period Duration of the part of a parasite's life cycle that is completed in a vector, that is the time from a vector becoming infected to it being infective (i.e. capable of transmitting the parasite).

Furuncular myiasis Dipterous larvae living in boil-like swellings, e.g. larvae of the tumbu fly.

Genetic control Special type of biological control that uses genetical techniques to control pest populations. Specifically the reduction of the reproductive potential of vectors. For example, the release of sterilized male vectors into field populations which results in producing large numbers of mated, but infertile, females. These females lay eggs that cannot hatch, so causing a population decrease. *See* Sterile male release.

Gonotrophic cycle Time from first blood-feeding to oviposition, and subsequently between successive ovipositions. The first such gonotrophic cycle may be a day or two longer than subsequent ones. Sometimes the gonotrophic cycle is defined as time from blood-feeding to blood-feeding. This cycle is sometimes called the ovarian cycle.

Habitat Usually means the physical environment in which an animal lives, e.g. the skin in the case of scabies mites, streams for simuliid larvae and animal nests for many ixodid ticks.

Haemocoel Main body cavity of arthropods in which insect blood (haemolymph) circulates.

Haemolymph Insect blood, usually colourless and clear, which fills the haemocoel.

Hemelytron (pl. hemelytra) Fore-wing of certain insects, e.g. triatomine bugs, which has a thickened basal part and a membranous distal part.

Hemimetabolous (incomplete metamorphosis) Describes the development from egg to adult which is gradual, passing through one or more nymphal stages, e.g. lice and ticks. If adults are winged the nymphal wing buds grow larger at each moult, e.g. bedbugs. There is no pupal stage. *See* Holometabolous.

Hereditary transmission Involves a female vector passing disease organisms to her eggs and thus to the next generation, i.e. trans-ovarial transmission.

Hibernation Period of inactivity and/or altered behaviour caused by cold conditions, e.g. winter. Some mosquitoes, e.g. *Culex pipiens* in Europe, may remain in complete hibernation without blood-feeding for many months by living off their fat reserves. Other species, e.g. *Anopheles atroparvus*, enter incomplete hibernation during which time adults need to emerge periodically from hibernation sites to take blood-meals to renew their fat reserves.

Holarctic Sometimes used to encompass the Palaearctic and Nearctic regions. *See* Zoogeographical regions.

Holometabolous (complete metamorphosis) Arthropod development from egg to adult in which the body form changes completely in appearance. A pupal or puparial stage is characteristic of holometabolous development. *See* Hemimetabolous.

Holoptic A condition of the head of adult Diptera in which the eyes meet or nearly meet each other, as, for example, in male adult simuliids. *See* Dichoptic.

Host preference Preferred hosts (e.g. species, sex, age) of an arthropod in an area when a choice exists. For a particular species this can alter from area to area as well as seasonally, depending usually on the availability of alternative hosts.

IGRs (insect growth regulators) Sometimes known as insect development inhibitors, these are groups of chemicals that either prevent the development of larvae into pupae or pupae into adults (juvenile hormone analogues, e.g. methoprene) or interfere with the moulting process killing larvae as they moult (chitin synthesis inhibitors, e.g. diflubenzuron).

Impoundment Well-defined man-made area of water, usually with more or less vertical sides, often constructed to provide water for domestic, agricultural or recreational puposes. However, impoundments are dug sometimes principally to reduce mosquito breeding in formerly marshy areas.

Infected Applied to arthropods when a parasitic infection has been taken up by the vector but is not yet in a stage in which it can be transmitted to a host. Examples are simuliid adults with onchocercal larvae in their flight muscles, or a vector with arboviruses in the stomach but not yet in the salivary glands. *See* Infective.

Infective Applied to arthropods when the parasites can be transmitted by the vector to a host. Examples are mosquitoes with malarial sporozoites in the salivary glands or simuliids with third-stage onchocercal worms in their mouthparts. *See* Infected.

Insecticide resistance Ability of arthropods to tolerate doses of insecticide which would prove lethal to the majority of normal (susceptible) individuals of the same species. Rare mutants which are resistant are selected for by the use of the insecticide.

Instar One of a series of life cycle stages in metamorphosis that is separated by a moult, e.g. the first, second and third larval instars of house-flies and the five nymphal stages of bedbugs.

Integument The cellular epidermis and outer non-cellular cuticle which together provide the outer covering of arthropods. *See* Exoskeleton.

Intermediate host Host in which a parasite does not reach sexual maturity. Applies to most parasites in arthropod vectors, e.g. filarial parasites in mosquitoes and simuliids. *See* Definitive host.

Intrinsic incubation period Duration of the life cycle of a parasite in the vertebrate host; interval between infection and first clinical symptoms.

Larviparous Reproduction in which the egg(s) hatch within the female and larva(e) are deposited, e.g. tsetse-flies.

Life cycle (life history) In entomology and parasitology this usually means the series of morphological stages an organism passes through to reach the mature adult stage and the biology of each stage.

Longevity How long an organism lives, often expressed as the mean expectancy of life. Vector longevity is one of the most important factors in disease transmission dynamics and vector control.

Maggot Legless larva that has no distinct head, thorax or abdomen, e.g. larvae of house-flies.

Maintenance host A vertebrate or arthropod host which allows long-term survival of parasite populations. The host must have an infection rate that is at least adequate to maintain a population of the disease agent continuously endemic in an area. Humans can be maintenance hosts of louse-borne typhus, and mosquitoes and birds appear to be maintenance hosts of some of the encephalitis viruses. A maintenance host may retain an infection during periods when there is no or little vector transmission, e.g. during dry seasons or winters.

Mechanical transmission Transmission where there is no multiplication or cyclical development of the aetiological agent (i.e. parasite or pathogen), it being merely passively carried by the vector. Examples are house-flies transmitting trachoma virus and dysenteries by their feet, vomit or faeces, and trypanosomes transmitted by stable-flies (*Stomoxys* species).

Metacyclic trypanosome (metatrypanosome) The final, and usually smaller, version of the trypomastigote form in the vector that is infective for the vertebrate host.

Metamorphosis Changes in form from the first stage (egg) in the life cycle of an arthropod to the adult form. In hemimetabolous arthropods, e.g. bedbugs, lice, ticks and mites, the change is gradual through nymphal stages which resemble small versions of the adult. In holometabolous arthropods, e.g. mosquitoes, tsetse-flies and fleas, the changes are abrupt and involve larval stage(s) and a pupal or puparial stage which are very dissimilar to the adult.

Microbial insecticide Insecticide comprising a biological agent such as bacteria, e.g. *Bacillus thuringiensis* subsp. *israelensis* and *B. sphaericus*, or toxic compounds derived from such agents.

Monolayer Thin film of a water-insoluble surfactant, e.g. lecithins, that has a very high spreading power on water. Mosquito larvae are unable to maintain normal contact with the water surface, and this combined with the reduction of dissolved oxygen content caused by monoloyers causes their death.

Morphology The outward structure of an organism. Most arthropods are identified by their morphology, that is by their outer appearance.

Moulting Process of shedding the cuticle between developmental stages (i.e. instars).

Myiasis Invasion of vertebrate organs or tissues by larvae of Diptera that feed on living or dead tissues.

Nearctic region The USA, Canada, Greenland and northern Mexico. *See* Zoogeographical regions.

Neotropical region South America, Central America, southern Mexico and Caribbean islands. *See* Zoogeographical regions.

New World North, Central and South America, and the Caribbean area. *See* Old World.

Nocturnal Refers to activity during the night, such as blood-feeding in anophelines and appearance in vertebrate blood of nocturnally periodic microfilariae of some helminths, e.g. *Wuchereria bancrofti*.

Nymph In incomplete metamorphosis (hemimetabolous development) the stage in the life cycle that hatches from the egg (e.g. bedbugs, lice), or the stage that arises through the moulting of the larval stage (e.g. ticks).

Old World All countries and areas east of the Americas. *See* New World.

Oocyst rate Percentage of mosquitoes that have malarial oocysts on the stomach.

Organochlorines (chlorinated hydrocarbons) Synthetic insecticides containing carbon, chlorine and hydrogen, e.g. DDT, dieldrin, HCH and methoxychlor.

Organophosphates Synthetic insecticides which are derivatives of phosphoric acid, hence all contain phosphorus, e.g. diazinon, dichlorvos and malathion.

Oriental region Asia east of Pakistan and south of the Himalayas and central China, Taiwan, Sri Lanka, and the South-east Asia archipelago eastwards to Sulawesi. *See* Zoogeographical regions.

Ornithophagic (ornithophilic) Arthropods that blood-feed on birds.

Osmoregulation The regulation of water balance in arthropods; maintaining the homeostasis (balance) of osmotic and ionic content of the body fluids.

Overwintering Describes the survival tactics of arthropods during winters. For instance adults (e.g. some mosquitoes) may cease feeding and ovipositing and enter a state of hiberation until warmer weather reappears and activity resumes. The growth and development of the immature stages (e.g. mosquito and simuliid larvae) may also slow down.

Palaearctic region Europe, North Africa, Asia north of the Himalayas and central China, Japan, Iceland, mid-Atlantic islands. *See* Zoogeographical regions.

Peritrophic membrane A thin tubular sheath secreted either by cells at the anterior end of the mid-gut (mosquito larvae, tsetse-fly adults) or by cells lining the mid-gut (adult mosquitoes). It is found only in some insects. It supposedly forms a protective lining for the mid-gut, but its exact function remains largely unknown. In tsetse-flies the peritrophic membrane plays a role in the cyclical development of human sleeping sickness trypanosomes.

Pheromone Chemical (semiochemical) released usually as an odour by an individual which produces reactions in others of the same species, e.g. sex pheromones in tsetse-flies and ticks.

Phoresy Transport of an animal from one place to another by means of attachment to another animal, e.g. *Simulium neavei* larvae on freshwater crabs, and eggs of *Dermatobia* flies attached to adult female mosquitoes.

Polytene chromosomes So-called giant chromosomes found only in certain tissues of Diptera, such as the ovarian nurse cells of half-gravid anophelines and larval salivary glands of simuliids. When stained they show distinct banding patterns which can often be used to identify species within species complexes. *See* Sibling species.

Promastigote (leptomonad) Morphological form of a trypanosomatid with the flagellum arising near the anterior end, e.g. *Leishmania* parasites in the phlebotomine sandfly gut.

Pseudopod (false leg) Stumpy protuberances present on dipterous larvae of some species (e.g. tabanids, phlebotomine sandflies) that assist them in locomotion. (None of the larvae of the Diptera possesses true legs.)

Pseudotracheae Small tubes in the labella of the adults of some Diptera (e.g. muscids, calliphorids and tabanids) which are supported by sclerotized rings. Liquid food passes through minute openings in these pseudotracheae to the mouth of the fly.

Puparium Life-stage resulting from the hardening and sclerotization of the cuticle of the last larval instar of certain Diptera, e.g. tsetse-flies and house-flies. Equivalent to the pupa of other insects.

Pyrethroids (synthetic pyrethroids) Synthetic insecticides containing different pyrethrin-like chemicals, e.g. permethrin, deltamethrin, lambdacyhalothrin.

Questing The behaviour of ticks, mainly ixodids, when climbing up vegetation, such as grasses and herbaceous plants, in order to actively seek out passing hosts, to which they attach themselves.

Quiescence Temporary state of arrested or slowed development, such as in ixodid larvae after blood-feeding but prior to moulting to the nymphal stage, or state of inactivity of some hibernating adult mosquitoes.

Reservoir Host animals in which populations of disease organisms persist indefinitely, and which pass the disease to other species of hosts, often by vectors. Reservoir hosts may be maintenance hosts, and are often mammals and birds. Some arthropod vectors which are long-lived, and may also be capable of trans-ovarial transmission, are sometimes regarded as reservoir hosts, e.g. ticks as vectors of relapsing fever and various viruses.

Residual spraying Application of a persistent insecticide (e.g. malathion) to surfaces, such as to the inside walls and roofs of houses in malaria control programmes, and to trees in tsetse control projects.

Rickettsiae A group of Gram-negative intracellular coccoid-shaped bacteria, many of which are transmitted by arthropods, e.g. *Rickettsia prowazeki* transmitted by body lice and causing typhus. Formerly rickettsiae were regarded as micro-organisms intermediate between bacteria and viruses.

Seed tick Name often given to the very small larva of an ixodid tick before it has blood-fed.

Sclerotization Process which results in the new arthropod cuticle formed after moulting being tanned (darkened) and hardened to give it rigidity.

Sibling species (isomorphic or cryptic species) Species that are morphologically indistinguishable or almost so, and which are recognized by non-morphological features, usually their polytene chromosomes (e.g. species of the *Simulium damnosum* complex). In nature they are reproductively isolated from each other. Sibling species may differ only slightly biologically but be significantly different in epidemiological importance.

s. l. (sensu lato) Means in the broad sense. When placed after the name of a species denotes that reference is made not just to that species but also to closely related species (e.g. sibling species) within a complex. *See s. str.*

Source reduction Simple measures that either prevent breeding of arthropods or eliminate their breeding sites. Mainly applicable to mosquito control, e.g. covering water tanks, filling in puddles or removing discarded water-retaining receptacles.

Species A group of individuals in natural populations that can interbreed by mating within the group and producing fertile progeny; individuals are usually similar in appearance and behaviour.

Species complex A group of sibling, or very closely related, species that are morphologically indistinguishable (isomorphic) but are reproductively isolated, and which often live in the same area (sympatric). Species within a complex can often be identified by their polytene chromosomes or by biochemical or molecular techniques. The Diptera contain many complexes, such as in mosquitoes (e.g. *Anopheles gambiae* complex) and simuliids (e.g. *Simulium damnosum* complex).

Species group Used for an assemblage of closely related species that although they may be morphologically similar in one or more life-stages can nevertheless be distinguished on external appearance as distinct species, e.g. the *Simulium neavei* group – comprising eight species including *S. neavei*, *S. woodi* and *S. nyasalandicum*.

Species sanitation Control, or eradication, directed against just one species of vector or pest in a particular area, usually using simple techniques.

Spermatheca One or more receptacles in the female reproductive system of arthropods which receive sperm during mating.

Spirochaetes Gram-negative bacteria which have a more or less spiral shape, e.g. *Borrelia duttoni* which is spread by soft ticks and causes relapsing fever.

Sporogony That part of the sexual cycle of sporozoans (e.g. *Plasmodium* species) in which sporozoites are produced.

Sporozoite rate Percentage of mosquitoes with malarial sporozoites in their salivary glands.

s. str. (sensu stricto) Means in the strict, or narrow, sense. When placed after a species name denotes that only that species is being referred to, and not any of its closely related (e.g. sibling) species in a complex. Sometimes abbreviated to *s. s. See s. l.*

Sterile male release (sterile-insect technique – SIT) In genetic control programmes the inundative release of large numbers of artificially sterilized male arthropods into field populations in the hope this will result in sterile matings and consequently a reduction in population size.

Subperiodic periodicity As applied to microfilariae in peripheral vertebrate blood means they exhibit partial diel (24 hour) periodicity. That is, their concentration in the blood decreases from a maximum to a minimum, but not close to zero as with microfilariae showing periodic periodicity. Diurnal subperiodicity occurs in the Pacific strain of *Wuchereria bancrofti* while nocturnal subperiodicity is found in populations of *Brugia malayi* in West Malaysia, Thailand, etc.

Subspecies In zoology the only taxon recognized below the rank of species. Subspecies differ morphologically from other members of the species and are spatially isolated, e.g. either geographically or by their hosts. Consequently they are normally reproductively isolated, but when brought together can interbreed with other subspecies. Subspecies may eventually evolve into distinct species.

Sylvatic In epidemiology means that diseases are contracted in woods or forests, e.g. the forest cycle of yellow fever. *See* Campestral.

Synanthropic Applied to animals living in close association with humans or their houses, e.g. house-flies and triatomine bugs.

Synergist Chemical that has little or no toxicity but, when combined with some insecticides, enhances their activity and thus reduces dosage rates. For example, piperonyl butoxide is a synergist added to the insecticide pyrethrum.

Trans-ovarial transmission Production by an infected vector of eggs infected with parasites (e.g. viruses, rickettsiae). When they hatch they give rise to individuals (e.g. larvae, nymphs) that are infected and are either capable of transmitting the parasites (e.g. ticks) or passing it on to later life cycle stages that transmit the infection (e.g. scrub typhus mites). *See* Hereditary transmission.

Trans-stadial transmission Survival of parasites through successive life cycle stages (larva – nymph(s) – adult) of an organism (e.g. ticks), each of which can transmit the parasite if it is haematophagous.

Traumatic myiasis The infestation of wounds by dipterous larvae, e.g. screwworm larvae. *See* Furuncular myiasis.

Trypomastigote Morphological form of a trypanosome with the flagellum arising near the posterior end, and running the length of the body where it is attached to the cell membrane. Trypomastigotes are found in the vertebrate blood of hosts infected with trypanosomiasis, and are the form ingested by a vector with its blood-meal.

Ultra-low-volume (ULV) Refers to spraying an insecticide in concentrated form; thus the dosage rate is small, e.g.<5 litres/ha.

Vector Organism that conveys an aetiological agent from one host to another. A vector may be an intermediate host (e.g. anophelines transmitting malaria) or not (e.g. house-flies mechanically transmitting bacteria).

Viraemia Presence of virus in vertebrate blood. High viraemias are usually necessary for transmission by arthropod vectors.

Wettable powder (water-dispersible powder) Technical grade insecticide diluted with an inert carrier (dust) and to which a wetting agent or surfactant has been added. The resultant wettable powder is then mixed with water for spraying onto surfaces.

Zoogeographical regions Any one of the six main geographical areas referred to by zoologists. Each region has its own particular fauna, of which many species occur only in that region. The Palaearctic and Nearctic region are sometimes grouped together and referred to as the Holarctic region. *See* Afrotropical, Australasian, Nearctic, Neotropical, Oriental and Palaearctic regions.

Zoonosis Natural transmission of infections between vertebrate hosts and humans.

Zoophagic (zoophilic) Blood-sucking arthropods that feed on non-human animals.

Index